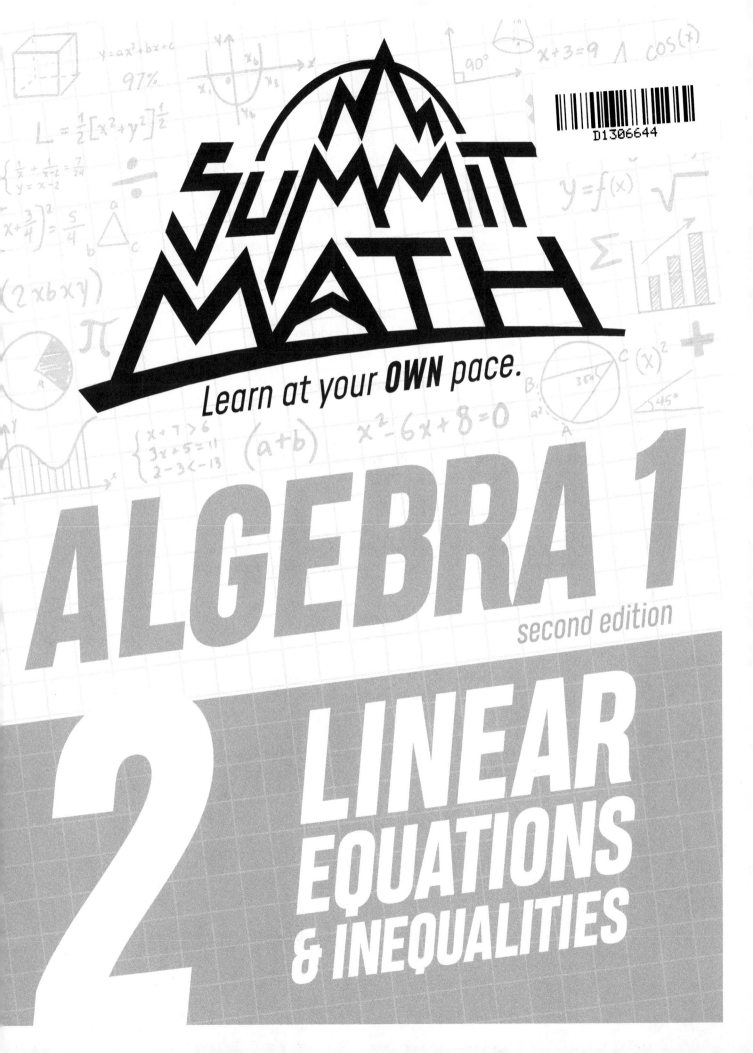

SUMMIT MATH

Learn at your **OWN** pace.

ALGEBRA 1

second edition

2 LINEAR EQUATIONS & INEQUALITIES

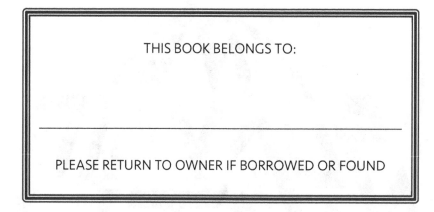

THIS BOOK BELONGS TO:

_____

PLEASE RETURN TO OWNER IF BORROWED OR FOUND

DEDICATION
To Lauren, Chloe, Dawson and Teagan

ACKNOWLEDGEMENTS
I started writing these books in 2013 to help my students learn better. I kept writing them because I received encouraging feedback from students, parents and teachers. Thank you to all who have used these books, pointed out my mistakes, and made suggestions along the way. Thank you to all of the students and parents who asked me to keep writing more books. Thank you to my family for supporting me through every step of this journey.

This book was typeset in the following fonts:
Seravek + Mohave + *Heading Pro*

Graphics in Summit Math books are made using the following resources:
Microsoft Excel | Microsoft Word | Desmos | Geogebra | Adobe Illustrator

First printed in 2017

Printed in the U.S.A.

Summit Math Books are written by Alex Joujan.

www.summitmathbooks.com

INTRODUCTION

Learning math through Guided Discovery:
A Guided Discovery learning experience is designed to help you experience a feeling of discovery as you learn each new topic.

Why this curriculum series is named Summit Math:
Learning through Guided Discovery can be compared to climbing a mountain. Climbing and learning both require effort and persistence. In both activities, people naturally move at different paces, but they can reach the summit if they keep moving forward. Whether you race rapidly through these books or step slowly through each scenario, this curriculum is designed to keep advancing your learning until you reach the end of the book.

Guided Discovery Scenarios:
The Guided Discovery Scenarios in this book are written and arranged to show you that new math concepts are related to previous concepts you have already learned. Try to fully understand each scenario before moving on to the next one. To do this, try the scenario on your own first, check your answer when you finish, and then fix any mistakes, if needed. Making mistakes and struggling are essential parts of the learning process.

Homework and Extra Practice Scenarios:
After you complete the scenarios in each Guided Discovery section, you may think you know those topics well, but over time, you will forget what you have learned. Extra practice will help you develop better retention of each topic. Use the Homework and Extra Practice Scenarios to improve your understanding and to increase your ability to retain what you have learned.

The Answer Key:
The Answer Key is included to promote learning. When you finish a scenario, you can get immediate feedback. When the Answer Key is not enough to help you fully understand a scenario, you should try to get additional guidance from another student or a teacher.

Star symbols:
Scenarios marked with a star symbol ★ can be used to provide you with additional challenges. Star scenarios are like detours on a hiking trail. They take more time, but you may enjoy the experience. If you skip scenarios marked with a star, you will still learn the core concepts of the book.

To learn more about Summit Math and to see more resources:
Visit www.summitmathbooks.com.

GUIDED DISCOVERY SCENARIOS

As you complete scenarios in this part of the book, follow the steps below.

Step 1: Try the scenario.
Read through the scenario on your own or with other classmates. Examine the information carefully. Try to use what you already know to complete the scenario. Be willing to struggle.

Step 2: Check the Answer Key.
When you look at the Answer Key, it will help you see if you fully understand the math concepts involved in that scenario. It may teach you something new. It may show you that you need guidance from someone else.

Step 3: Fix your mistakes, if needed.
If there is something in the scenario that you do not fully understand, do something to help you understand it better. Go back through your work and try to find and fix your errors. Mistakes provide an opportunity to learn. If you need extra guidance, get help from another student or a teacher.

After Step 3, go to the next scenario and repeat this 3-step cycle.

NEED EXTRA HELP?
watch videos online

Teaching videos for every scenario in the Guided Discovery section of this book are available at www.summitmathbooks.com/algebra-1-videos.

CONTENTS

Section 1
PLOTTING POINTS ON A GRAPH

1. As part of a health test, a doctor wants to monitor your heart rate. At 2:00pm, the doctor fastens a heart rate monitor onto your arm and then asks you to walk over to a treadmill and start jogging. After a couple of minutes, your results are displayed on the graph, as shown.

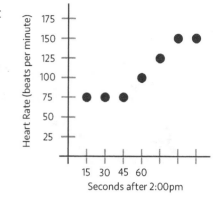

   a. Looking at the points from left to right, what information does the <u>first</u> point display?

   b. Looking at the points from left to right, what information does the <u>last</u> point display?

2. On a cold day in January, the temperature barely reaches 15° Fahrenheit (°F) at 5:00pm before getting colder again as the sun starts to set.

   Looking at the graph shown below, what information is displayed by the 3 points?

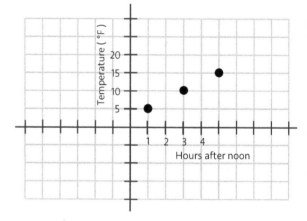

3. Place more points in the graph to represent each of the following statements.

   a. At 7:00pm, the temperature was 10°F.

   b. At 9:00am, it was 5° below zero.

   c. At 7:00am, it was 10° below zero.

   d. At 9:00pm, it was 5°F.

4. In the scenario above, it becomes useful to add more marked numbers on the number lines, including values that are negative to represent temperatures <u>below</u> 0° and hours <u>before</u> noon. At what rate did the temperature increase from 1:00pm to 5:00pm, measured in degrees per hour?

The previous graphs are simplified to help you focus on a small set of information as you begin learning about graphing. In both examples, you can "read" the information contained in the graphs by looking at the location of the dots. Each graph contains a numbered horizontal and vertical axis with labels to describe what the numbers mean.

In the 17th century (the 1600s), a French philosopher named René Descartes (day-CART) created a simple way to compare two numerical quantities. He realized that he could take two number lines and make them intersect at 90° angles, essentially forming a plus sign (+). In honor of his discovery, this structure is called a Cartesian plane, although it is also often called a "graph."

5. In the graph shown, draw a dot 5 units to the right of the vertical axis and 3 units above the horizontal axis.

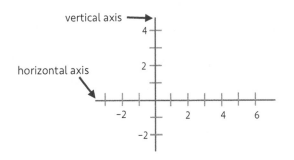

6. In the graph shown, draw a dot 3 units to the left of the vertical axis and 1 unit above the horizontal axis.

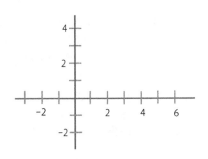

The description of each dot's location requires an entire sentence, so mathematicians have created a way to express the location more concisely. In the graphs above, the location of each dot is usually expressed as a pair of numbers, separated by a comma, enclosed by parentheses.

In the first graph, the location of the dot is written as $(5, 3)$. In the second graph, the location of the dot is $(-3, 1)$. There is a defined order to the numbers. The numbers in parentheses will always be written as (horizontal location, vertical location). As a result, $(5, 3)$ is called an ordered pair. It may also be called a "coordinate" or a "point."

7. Describe the location of each ordered pair shown. Then mark each ordered pair in the graph.

DESCRIPTION: From the origin, move. . .

a. $(2, 7)$ right 2 units, up 7 units

b. $(-3, 6)$

c. $(5, -1)$

d. $(-7, -2)$

e. $(0, 3)$

f. $(-4, 0)$

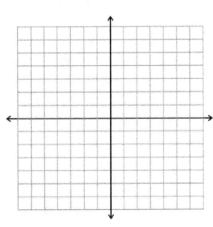

8. Place the points in their correct location on the graph. Label each point with its corresponding letter.

a. $(2, 3)$ b. $(-4, -1)$

c. $(-3, 2)$ d. $(-6, 0)$

e. $(3, -1)$ f. $(0, -3)$

g. $(6, 1)$ h. $(0, 4)$

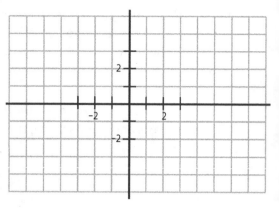

9. It is easier to work with small numbers, but plotting points with larger numbers is worth practicing because it forces you to estimate. You may not be exact, but you can get close. As closely as you can, estimate the location of each point below. Note how the axes are marked.

a. $(25, -300)$ b. $(7.5, -250)$

c. $(-5, 0)$ d. $(-27.5, 225)$

e. $(0, -325)$ f. $(-16, -60)$

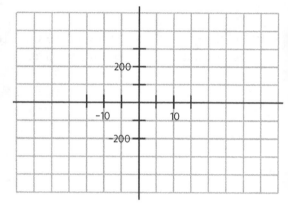

10. A typical bathtub faucet has a flow rate of 2 gallons per minute. At 8:00pm, a faucet is turned on and a bathtub begins filling with water.

a. How much water is in the bathtub at 8:01pm?

b. At 8:07pm?

c. At 8:13pm?

d. Draw points in the graph to show the change in the amount of water in the bathtub. Number your horizontal axis (m-axis) and vertical axis (A-axis) before you draw the points.

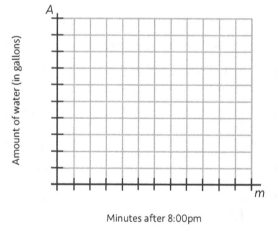

Minutes after 8:00pm

11. In the previous scenario, write an equation that relates the amount of water in the bathtub, A, to the number of minutes, m, that have passed since 8:00pm.

12. A garden hose has a different flow rate than a bathtub faucet. When you fill a swimming pool with a garden hose, the amount of water in the pool, in gallons, is modeled by the equation $A = 6m$, where A is the amount of water in the pool after the hose has been pumping water for m minutes.

   a. What is the flow rate of this garden hose? Include correct units in your response.

   b. How much water would be in the pool after 5 minutes?

   c. How much water would be in the pool after 6 minutes and 30 seconds?

13. Now make the equation $A = \dfrac{3}{2}m + 5$.

   a. What is the value of A when $m = 12$?

   b. What is the value of A when $m = 0$?

14. Now change the previous equation to $y = -3x + 11$.

   a. What is the value of y when $x = 2$?

   b. What is y when $x = 0$?

NOTES

Use this page to record important ideas in the previous section or
for any other writing that helps you learn the topics in this book.

Section 2
GRAPHING A LINE USING AN EQUATION & A T-CHART

15. Consider the equation $y = x - 4$.

    a. If x is 0, what is y?
    b. If $x = 1$, the y-value is ____.
    c. If $x = 2$, y is ____.

    d. If $x = 3$, y is ____.
    e. If $x = -1$, $y =$ ____.
    f. If $x = -2$, $y =$ ____.

16. In the previous scenario, each pair of numbers can be written as an ordered pair, (x, y). Write each of the ordered pairs in the space below.

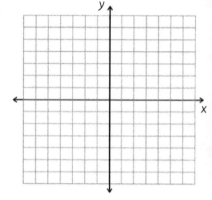

17. If each x-value is a horizontal location, and each y-value is a vertical location, plot the previous ordered pairs in the graph to the right.

18. When you plot the points for the equation in the previous scenario, what do you notice about the way they appear in the graph?

In the previous scenario, you find ordered pairs by plugging x-values into the equation. When you want to find ordered pairs for an equation, you can organize your work using a chart with 2 columns, sometimes called a T-chart.

19. Graph the equation $y = -x + 3$ by using a T-chart. First, fill in the missing values in the T-chart. Second, write each pair of x- and y-values as an ordered pair. Finally, plot the points in the graph.

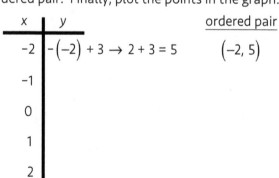

x	y		ordered pair
-2	$-(-2) + 3 \to 2 + 3 = 5$		$(-2, 5)$
-1			
0			
1			
2			

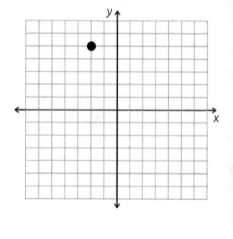

20. When you plot the points for the equation in the previous scenario, what do you notice about the way they appear in the graph?

21. Graph the equation $y = 2x - 3$ by using a T-chart.

x	y		ordered pair
-2	$2(-2) - 3 \rightarrow -7$		$(-2, -7)$
-1			
0			
1			
2			

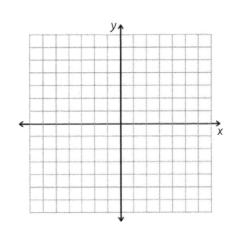

22. Graph the equation $y = -\dfrac{1}{2}x + 4$ using a T-chart.

a. This time, only find four ordered pairs using the chart, unless you want to find more.

b. Without using the T-chart, how could you graph 100 more points in less than 5 seconds?

c. What is the y-value if x is −10?

d. What is the x-value if y is −2?

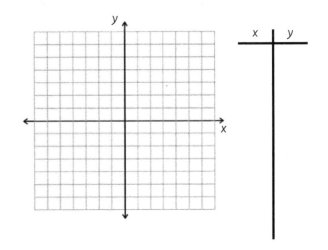

x	y

23. Graph $6x + 3y = 3$ using a T-chart. To save time, only graph 3 ordered pairs. Draw a line through your points to show that there are more points than the three you found.

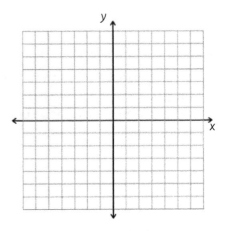

24. Consider the line in the previous scenario.

   a. How could you determine if $(1.5, -1)$ is located exactly on the line?

   b. Is $(-1.1, 3.2)$ on the line?   c. Is $(201, -400)$ on the line?

25. Like the previous scenarios, the equation $2x - 5y = 10$ forms a line when you draw its graph. Now that you know this, graph the points below and try to guess which ones are <u>NOT</u> on this line.

   a. $(5, 0)$ b. $(-2, 4)$ c. $\left(\dfrac{5}{2}, -1\right)$

   d. $(6, 3)$ e. $(0, -2)$ f. $(-5, -4)$

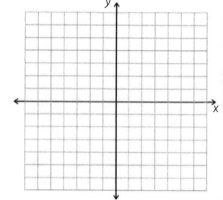

26. The equation $4x + 2y = 16$ also forms a line when you draw its graph. How can you find points that are on the line $4x + 2y = 16$? Do <u>not</u> actually find points on this line.

27. When you make a T-chart to find points on the line $4x + 2y = 16$, you can start with an x-value and find the y-value that goes with it. Together these make an ordered pair. Read each statement below and write the information as an ordered pair.

   a. If $x = 4$, $y = 7$.   b. If $x = -5$, $y = 23$.   c. When $y = -2$, $x = 11$.

28. When you plug numbers into an equation, some numbers are easier to work with. What is the easiest number to plug into an equation?

NOTES

Use this page to record important ideas in the previous section or
for any other writing that helps you learn the topics in this book.

Section 3
GRAPHING A LINE USING ITS INTERCEPTS

In the previous scenario, you were asked to think about a number that is easy to work with. In this section, you will learn how to graph an equation by replacing each variable with a "0".

29. Consider the equation $-2x + 4y = 8$.

    a. Replace x with 0 and solve for y.

    b. Start with the original equation again. This time, replace y with 0 and solve for x.

    c. You now know 2 points on the line $-2x + 4y = 8$. Write the two points as ordered pairs.

30. Graph the points you found in the previous scenario and draw a line through them. After you draw the line, try to guess some other points that are on the line and plot them also.

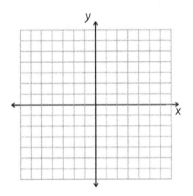

31. Graph $3x + 6y = 18$ by repeating the strategy that you used in the previous scenario.

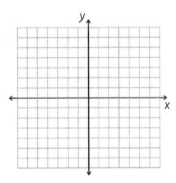

32. If a point has an <u>x-value of 0</u>, where will that point be located on the Cartesian plane?

33. If a point has a <u>y-value of 0</u>, where will it be located on the Cartesian plane?

If you graph a line on the Cartesian plane and the line crosses the x-axis, the point where the line touches the axis is called an **x**-intercept. Similarly, if a line crosses the y-axis, the point where it touches the axis is called a **y-intercept**.

34. Circle one of the terms to the right of the statement.

    a. If the coordinates of a point are $(K, 0)$, then it is a(n)   x-intercept   y-intercept

    b. If the coordinates of a point are $(0, K)$, then it is a(n)   x-intercept   y-intercept

35. Consider the graph shown to the right.

    a. Identify the x-intercept of the line.

    b. Identify the y-intercept of the line.

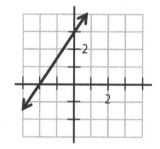

36. The equation of a line is 3x + 5y = 12. If you want to find the x-intercept of this line without graphing, how would you do this?

37. The equation of a line is 2x – 9y = 27. If you want to find the y-intercept of this line without graphing, how would you do this?

38. Find the coordinates of the x- and y-intercepts of the equation –11x + 4y = –22.

39. Graph 5x – 6y = 15 by finding the x- and y-intercept of the line. Graph these two points and then use the points to draw the line.

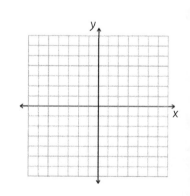

Many of the previous equations have the same structure: Ax + By = C, where A, B, and C are <u>integers</u>. Mathematicians refer to Ax + By = C as the Standard Form of a linear equation.

40. Circle the equations that are in Standard Form, where A, B, and C are <u>integers</u>.

a. $3x + 6y = 6$

b. $9x - 4y = -12$

c. $\frac{1}{2}x - 5y = 30$

d. $\frac{3}{4}x + 2y = \frac{1}{4}$

e. $x + y = 4$

f. $-3x + 2y = 7$

41. Two points for the graph of $-2x - 3y = 12$ are shown. Use the equation to find <u>two more points</u> and draw both of them on the graph. What do you notice?

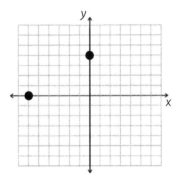

42. Write the Standard Form of a linear equation.

43. After a bathtub is filled, at some point it must be emptied out again. Let's assume a tub is filled with 21 gallons of water and at 8:30pm, the water begins to drain. Assume that the water flows down the drain at a rate of 1.5 gallons per minute.

a. Write an equation that shows the amount of water in the bathtub, A, after the tub has been draining for m minutes.

b. Check to see that your equation is accurate and then graph the equation.

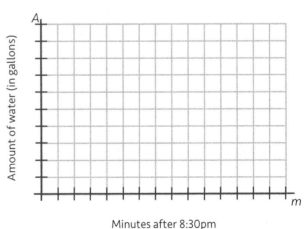

Minutes after 8:30pm

c. Identify the x- and y-intercepts of the line and explain what they represent in this scenario.

17

NOTES

Use this page to record important ideas in the previous section or for any other writing that helps you learn the topics in this book.

Section 4
CONSTANT RATES

Earlier scenarios already involved rates, but it is time to take another look at this topic.

44. Your grandmother opens a savings account for you and initially puts some money in your account. You set a goal to save up enough money to buy a new computer and you start setting aside the same amount of money every week. After 5 weeks, you have $320 in your savings account. After 11 weeks, you have $464. How much money are you setting aside each week?

In the previous scenario, the same amount of money is set aside every week. The constant increase every week allows you to predict the amount of money in the account at any given week. Consider some more scenarios involving constant rates.

45. Consider the graph to the right.

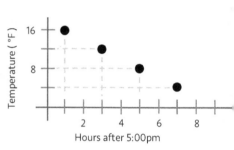

   a. What rate can be determined by analyzing the graph?

   ★b. Find the reciprocal of your rate in part a. and explain what it means.

46. Three families go to a blueberry farm and pick different amounts of blueberries. The amounts that they each picked are shown in the graph, as well as the amount that they each paid for their blueberries.

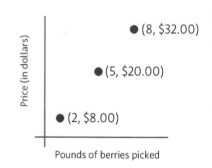

   a. How much did the blueberry farm charge for the berries? Express your answer as a rate.

   b. How many pounds of blueberries can someone buy for $1?

47. A portion of a graph is shown.

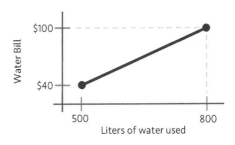

a. What type of rate can be found by analyzing the graph?

b. Write the coordinates of both endpoints of the segment.

c. How can you use the numbers in your ordered pairs to find the cost per liter?

d. What is the cost per liter?

48. ★Suppose your cell phone plan costs $80/month and allows you to send up to 500 text messages. Once you go over your limit, you are charged for each additional message that you send. The graph to the right displays the cost of your cell phone bill based on the number of messages sent in one month.

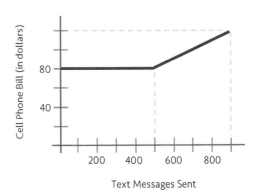

a. What is the cost per message after you exceed your monthly limit?

b. Find the reciprocal of your rate in part a. and explain what it means. Include units.

NOTES

Use this page to record important ideas in the previous section or
for any other writing that helps you learn the topics in this book.

Section 5
THE SLOPE OF A LINE

The Americans with Disabilities Act of 1990 (ADA) prohibits discrimination based on disability. One of the requirements of the ADA is that ramps for wheelchair usage may not be too steep. At what point does a ramp become too steep? To start with, define steepness as a ratio of the vertical and horizontal measurements of a ramp. As a fraction, steepness is $\dfrac{\text{vertical distance}}{\text{horizontal distance}}$.

49. The requirement created by the ADA is as follows: for every 1 foot that a ramp rises vertically, it should extend 12 feet horizontally. The required steepness, then, is 1:12. If you express this as a fraction, the required steepness is $\dfrac{1}{12}$.

    a. If a ramp extends 24 feet horizontally, what is the required height of the ramp above the ground?

    b. If a ramp extends horizontally 54 feet and vertically 4.5 feet, prove that it is legally constructed.

50. In the previous scenario, steepness is defined as "vertical distance over horizontal distance" but it will soon become tedious to keep referring to that definition. From here on, steepness will be replaced with the word <u>slope</u> and it will be defined as the fraction "rise over run." For example, consider the ramp shown below.

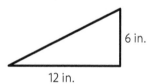

    a. The ramp rises ____ inches and it runs ____ inches.

    b. Thus, its $\dfrac{\text{rise}}{\text{run}}$ is ——— .

    c. If you simplify the slope, it can be written as ——— .

51. Identify the slope of each ramp shown.

a.
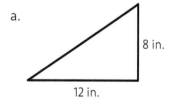

b.
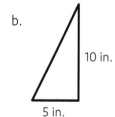

c.
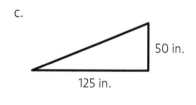

52. Identify the slope of each ramp shown.

a.

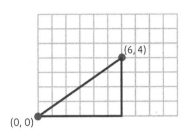

b.

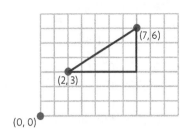

53. Identify the slope of each ramp shown.

a.

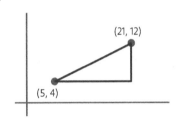

b.

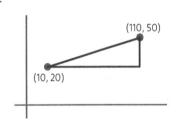

54. Let's move beyond ramps. Since every slanting line is like a ramp, we can take a section of any line and identify its slope, or its "rise over run." Identify the slope of the line shown and explain what the slope means using the labels for each axis.

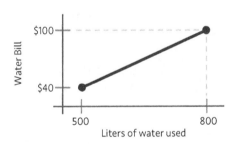

55. Identify the slope of the line shown and explain its meaning.

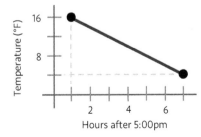

56. In the previous scenario, you are given a line that is sloping downward from left to right, after seeing lines that only sloped upward from left to right in previous scenarios. The two graphs shown represent lines that have a slope that can be expressed as $\frac{12}{6}$, or $\frac{2}{1}$, or simply 2. In both graphs, the slope of the line shows that the temperature is changing at a rate of 2 degrees per hour. If both graphs have the same slope, could they both describe the same scenario?

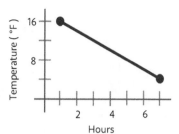
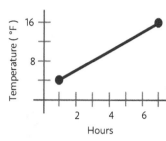

57. A line's slope is "rise over run." The rise can be up or down. The run can be left or right. To separate up from down and left from right, we need negative numbers. Since the Cartesian plane is 2 intersecting number lines, we can use these number lines to determine which directions are negative.

   a. Between up and down, which direction is a negative movement?

   b. Between left and right, which direction is a negative movement?

58. Two points are labeled in the graph to the right.

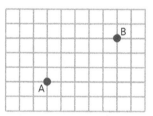

   a. Use the words up, down, left, and right to describe how to move from point A to point B in the graph shown.

   b. Use the words up, down, left, and right to describe how to move from point B to point A in the graph shown.

   c. Draw a line that passes through points A and B. What is the slope of this line?

59. Two points are labeled in the graph to the right.

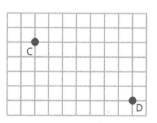

   a. Use the words up, down, left, and right to describe how to move from point C to point D in the graph shown.

   b. Use the words up, down, left, and right to describe how to move from point D to point C in the graph shown.

   c. Draw a line that passes through points C and D. What is the slope of this line?

60. Fill in the blanks. The slope of a line is _____ over _____. It is easy to forget the order of these two measurements. It will be helpful to practice writing this as a fraction in the right order.

 a. The slope of a line is $\dfrac{rise}{}$.

 b. The slope of a line is $\dfrac{}{}$.

61. The slope of a line is $\dfrac{2}{3}$. One of the points is shown in the graph. If you were standing on that point, where would you walk to get to another point on the line? Your directions must include whole number movements. Your directions must include the words up or down, and left or right.

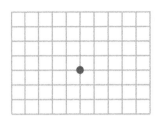

62. The slope of a line is $-\dfrac{2}{3}$. One of the points is shown in the graph. If you were standing on that point, where would you walk to get to another point on the line? Your directions must include whole number movements. Your directions must include the words up or down, and left or right.

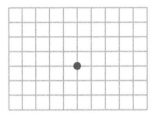

63. Draw the line that has the slope shown and passes through the point that is already plotted for you in the graph. Draw three more points before you draw the line.

 a. Slope: $\dfrac{2}{3}$

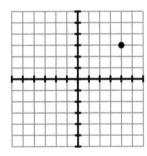

 b. Slope: $\dfrac{-1}{1}$

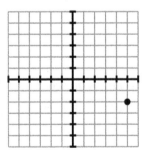

64. Every time you have found the slope of a line so far, two points have been marked for you. If you have more than two points marked, you will need to decide which two points you will use to identify the rise and the run of the line. In the graph, pick a pair of points and determine the slope of the line. Then pick a different pair of points and find the slope of the line again. Compare your slopes.

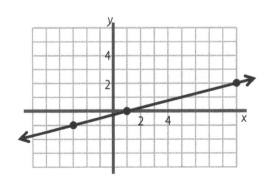

65. In the graph, pick a pair of points and determine the slope of the line. Then pick a different pair of points and find the slope of the line again. Compare your slopes.

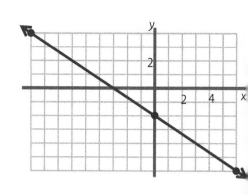

66. Identify the slope of the dashed line and the solid line in each graph below.

a.

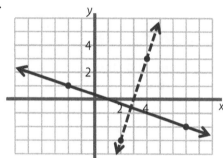

b.

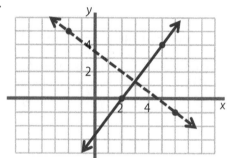

67. A line is drawn on the Cartesian plane but you only know two of the ordered pairs on the line. The ordered pairs are $(2, 6)$ and $(8, 10)$. Without graphing the points, identify the slope of the line.

68. Suppose another line is drawn and it passes through the ordered pairs $(-3, -7)$ and $(0, 2)$. What is the slope of the line?

69. Determine the slope of the line passing through each set of ordered pairs.

  a. $(3, -9), (-3, 6)$ b. $(-10, 2), (6, -8)$ ★c. $\left(\dfrac{7}{5}, 20\right), \left(\dfrac{1}{2}, 2\right)$

70. Mark two points on the dashed line to the right and use those points to find the slope of the line. What do you notice?

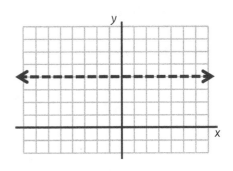

71. What is the slope of a horizontal line? Explain why a horizontal line has this particular slope.

72. Mark two points on the dashed line to the right and use those points to find the slope of the line. What do you notice?

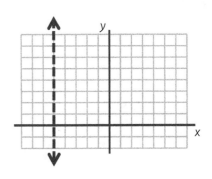

73. What is the slope of a vertical line? Explain why a vertical line has this particular slope.

74. Use the graph to answer each question.

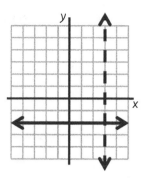

    a. What is the slope of the dashed line?

    b. What is the slope of the solid line?

75. Identify the slope of the line passing through each set of ordered pairs.

    a. $(-3, -6), (2, -6)$ b. $(9, -2), (9, 3)$

76. Write the Standard Form of a linear equation.

NOTES

Use this page to record important ideas in the previous section or
for any other writing that helps you learn the topics in this book.

Section 6
WRITING A LINE'S EQUATION IN SLOPE-INTERCEPT FORM

77. Let's take another look at graphing. Consider the equation $y = 3x - 5$.

    a. Fill in the T-chart and graph the three ordered pairs.

x	y
1	
2	
3	

    b. Plot three more points on the line.

    c. Use the points to find the slope of the line $y = 3x - 5$.

    d. Write the equation of the line again.

    e. Now write the slope of the line again.

78. Consider the equation $y = -2x + 4$.

    a. Fill in the T-chart and graph the three ordered pairs.

x	y
1	
2	
3	

    b. Plot five more points on the line.

    c. Use the points to find the slope of the line $y = -2x + 4$.

    d. Write the equation of the line again.

    e. Now write the slope of the line again.

79. What did you notice when you wrote the equation of the line and the slope of the line?

80. If the equation of a line is $y = 6x - 2$, what is the slope of this line?

81. If the equation of a line is $y = \frac{3}{4}x - 1$, what is the slope of this line?

82. Use what you have noticed in the previous scenarios to fill in the blanks below.

    a. The line formed by the equation $y = 4x - 7$ has a slope of _____.

    b. The line formed by the equation $y = \dfrac{1}{2}x + 2$ has a slope of _____.

    c. The line formed by the equation $y = 2x + \dfrac{1}{2}$ has a slope of _____.

    d. The line formed by the equation $y = Ax + B$ has a slope of _____.

83. Write an equation for a line that has a slope of $\dfrac{5}{11}$.

84. When a linear equation is arranged to look like $y = Ax + B$, the slope of the line will always be "A". What is the slope of the line formed by the equation $-5x + y = 10$?

85. What is the slope of the line that has an equation of $2x + 3y = 12$?

86. Let's take another look at the equation $y = 3x - 5$.

    a. Do you remember how to find the y-intercept of a line if you have the equation?

    b. Find the y-intercept of the line that is formed by the equation $y = 3x - 5$.

    c. Look at the numbers in the original equation. Do you notice anything?

87. Now consider the equation $y = -2x + 4$. Find the y-intercept of the line that is formed by this equation. Look at the numbers in the original equation. Do you notice anything?

88. What is the y-intercept of the line formed by the equation $y = \frac{2}{5}x - 8$?

89. The line formed by the equation $y = 5x - 31$ will cross the y-axis at (0, ____).

90. What are the coordinates of the y-intercept of the line formed by the equation $y = 5x$?

91. When a linear equation looks like $y = Ax + B$, the y-intercept will be located at "B," or (0, B). What are the coordinates of the y-intercept of the line formed by the equation $2x + 3y = 12$?

Since a linear equation with the structure $y = Ax + B$ shows you the line's <u>slope</u> and <u>y-intercept</u>, it is usually called the Slope-Intercept Form. However, since A and B are already used to describe an equation in Standard Form (Ax + By = C), the Slope-Intercept Form is usually referred to as $y = mx + b$. Whichever letters you use, it is more important to see the information contained in this equation.

92. Write the Slope–Intercept Form of a linear equation. Then write the Standard Form.

93. Consider the equation $y = \frac{1}{3}x - 2$. Locate the y-intercept and draw a point to show where it is on the graph. Now identify the slope and use it to graph another ordered pair. Continue using what you know about the slope to graph other ordered pairs until you have plotted a total of 5 points. Then quickly graph another seven hundred points.

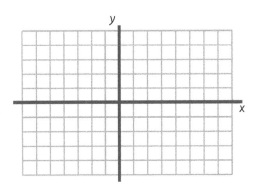

94. If the slope of a line is $\frac{1}{3}$, you can start at any point on the line and move up 1 unit and right 3 units to end up at another point on the line. If you start at a point on the line and move <u>down</u> 1 unit, how would you move horizontally to end up on the line again?

95. Suppose the slope of a line is $-\frac{2}{7}$. If you start at one point on the line, describe horizontal and vertical movements that would allow you to move to another point on the line.

96. The slope of a line is 2. One of its points is shown in the graph.

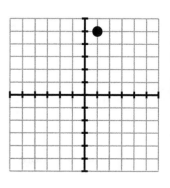

    a. Draw <u>three</u> more points that are on this line.

    b. The y-intercept of this line is located at (___ , ___).

    c. One of the points on this line is located at $\left(-5 , \underline{}\right)$.

97. The slope of a line is $-\dfrac{4}{5}$. One of its points is shown in the graph.

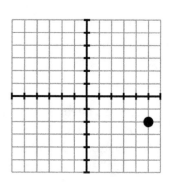

    a. Draw <u>two</u> more points that are on this line.

    b. The y-intercept of this line is located at (___ , ___).

    ★c. One of the points on this line is located at $\left(1 , \underline{}\right)$.

98. Graph the equation $y=-\dfrac{3}{2}x+1$ by locating the y-intercept as your first point and then using the slope to find 3 more points.

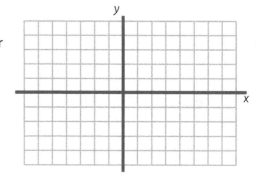

99. Circle the equation below that does <u>not</u> have the same graph as the other ones. Graph each equation on the same Cartesian Plane and compare the three graphs.

$$y=-\frac{1}{4}x+3 \qquad y=\frac{-1}{4}x+3 \qquad y=\frac{1}{-4}x+3$$

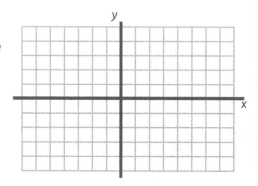

100. If the previous scenario is confusing, consider the following. What is the value of each fraction below, when it is simplified as much as possible?

    a. $\dfrac{-8}{4}$ b. $\dfrac{8}{-4}$ c. $-\dfrac{8}{4}$

101. Graph each line on the same plane.

    a. $y = 1x - 5$

    b. $y = 2x - 5$

    c. $y = 3x - 5$

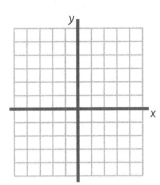

102. Graph the line given by each equation.

    a. $y = -\dfrac{2}{3}x + 4$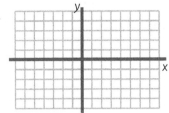

    b. $y = -x + 5$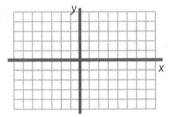

103. Try to graph the equation $4x + 6y = -6$, using the same approach that you have used in the previous scenarios. What makes this more challenging than graphing the previous equations?

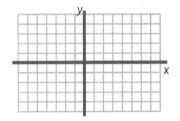

104. Each equation below is in Standard Form. Rewrite it to convert it to Slope-Intercept Form.

    a. $3x - 2y = 4$ b. $x + 2y = -14$ c. $3x + 3y = 15$

105. Which equation does <u>not</u> have the same graph as the other ones?

    $y = \dfrac{-2}{3}x + 5$ $y = \dfrac{4}{-6}x + 5$ $3y = -2x + 15$

106. What is the slope of a line if its equation is $y = -\frac{1}{5}x + 6$?

107. The equation $y = 5x - 6$ represents a line with a slope of 5. The equation $3y = 5x + 10$ also looks like it represents a line with a slope of 5.

    a. Why is this the incorrect slope for the line $3y = 5x + 10$?

    b. Determine the slope of the line $3y = 5x + 10$.

108. Graph the equations $-2x + 2y = 6$ and $3x - 3y = 12$ on the same Cartesian plane. What do you notice?

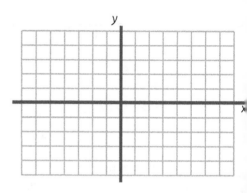

109. Fill in the blanks.

    a. The equation of the dashed line is $y = x -$ _____.

    b. The equation of the solid line is $y =$ ____$x + 1$.

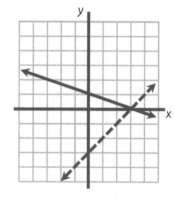

110. Write the equation for each line shown below. Write the equation in Slope-Intercept Form.

    a.

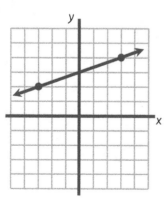

    b.

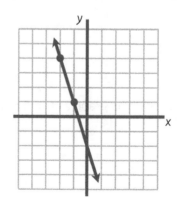

    c.

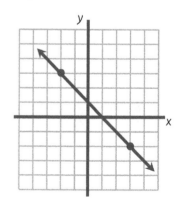

111. Which line passes through the point $(3, 7)$?

    Line 1: $y = 3x$ Line 2: $y = 3x - 1$ Line 3: $y = 3x - 2$

112. Like the line above, the line $y = 4x + b$ also contains the point $(3, 7)$. What is the value of b?

113. If a line has a slope of $\frac{2}{3}$, and a y-intercept of $(0, 4)$, it is easy to write the line's equation in Slope-Intercept Form ($y = mx + b$). Replace "m" with $\frac{2}{3}$. Replace "b" with 4. The equation is $y = \frac{2}{3}x + 4$.

    However, if only the slope is known, such as $\frac{2}{3}$, then only part of the equation is known: $y = \frac{2}{3}x + b$. You must _find_ the y-intercept in order to write the entire equation of the line. If you are told that the point $(6, 2)$ is located on the line, can you figure out a way to find the value of b?

114. What is the y-intercept of a line that has a slope of $-\frac{1}{5}$ and passes through the point $(5, 7)$?

115. What is the y-intercept of a line that has a slope of 5 and passes through the point $(17, 122)$?

116. There is some water in a tank, but more water needs to be added. When a valve is opened, water flows into the tank at a constant rate of 5 gallons per minute. After the valve has been open for 17 minutes, 122 gallons of water are in the tank.

    a. How many gallons of water were in the tank at the moment the valve was opened?

    b. In what ways is this scenario identical to the previous scenario?

117. If you know 2 points that are on a line, but you don't know the slope of the line, how would you find the equation of the line in Slope-Intercept Form?

118. Find the equation of the line that passes through the given points.

    a. $\left(-2,\ -5\right)$ and $\left(1,\ 7\right)$ b. $\left(9,\ 2\right)$ and $\left(-3,\ 10\right)$

119. When a valve is opened, water flows through a pipe at a constant rate into a tank that already contains some water. After the valve has been opened for 8 minutes, there are 72 gallons of water in the tank. After the valve has been opened for 21 minutes, there are 163 gallons of water in the tank.

    a. How fast is water flowing into the tank, measured in gallons per minute?

    b. How many gallons of water were in the tank at the moment the valve was opened?

120. Find the equation of the line that passes through the points $\left(8,\ 72\right)$ and $\left(21,\ 163\right)$.

121. Find the equation of the line shown.

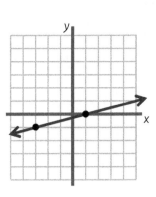

122. In the graph shown to the right, assume that ordered pairs that appear to be integers <u>are</u> integers.

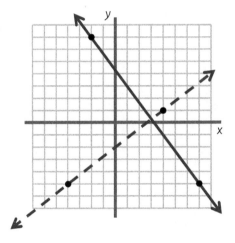

    a. Identify the equation of the solid line.

    b. Identify the equation of the dashed line.

NOTES

Use this page to record important ideas in the previous section or
for any other writing that helps you learn the topics in this book.

Section 7
PARALLEL & PERPENDICULAR LINES

123. Notice the slopes of the lines in the previous scenario. These lines are perpendicular. Make a guess about how you can see that lines are exactly perpendicular if you only know their slopes.

124. Find the slope of each line shown.

a.

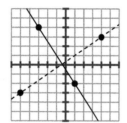

b.

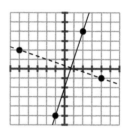

125. Both pairs of lines in the previous scenario are perpendicular. They intersect at right angles. Using what you have seen in the previous scenario, fill in the blanks below.

Perpendicular lines have slopes that are _____ _____.

126. Write the opposite reciprocal of each number shown below.

a. $\dfrac{2}{5}$

b. $-\dfrac{7}{4}$

c. $\dfrac{1}{3}$

d. -8

127. Fill in the box to make the two equations represent lines that are perpendicular.

a.
$$y = \dfrac{3}{5}x + 7$$
$$y = \boxed{}x - 2$$

b.
$$y = \boxed{}x - 1$$
$$y = -3x + 5$$

c.
$$y = -\dfrac{1}{7}x + \dfrac{3}{7}$$
$$y = \boxed{}x + 9$$

128. Fill in the box to make the two equations represent lines that are parallel.

a.
$$y = \frac{3}{5}x + 7$$
$$y = \boxed{}x - 2$$

b.
$$y = \boxed{}x - 1$$
$$y = -3x + 5$$

c.
$$y = -\frac{1}{7}x + \frac{3}{7}$$
$$y = \boxed{}x + 9$$

129. Do the equations represent parallel or perpendicular lines?

a.
$$y = \frac{1}{2}x - 2$$
$$y = \frac{1}{2}x + 7$$

b.
$$y = \frac{4}{3}x - 6$$
$$y = -\frac{4}{3}x - 2$$

c.
$$y = -\frac{5}{2}x + 11$$
$$y = 4 + \frac{2}{5}x$$

130. Do the equations represent parallel or perpendicular lines?

a.
$$3x + 7y = 21$$
$$-6x - 14y = 28$$

b.
$$30x + 40y = -400$$
$$-4x - 3y = 12$$

NOTES

Use this page to record important ideas in the previous section or
for any other writing that helps you learn the topics in this book.

Section 8
SCENARIOS THAT INVOLVE LINEAR EQUATIONS

131. Let's revisit an earlier scenario. Your grandmother opened a savings account for you and put some money in your account to help you get started. You set a goal to save up enough money to buy a new computer and started setting aside the same amount of money every week. After 5 weeks, you had $320 in your savings account. After 11 weeks, you had $464.

    a. Write an equation that represents the amount of money, A, in your savings account after you have been saving money for w weeks.

    b. How much money will you have in your savings account after 20 weeks?

132. Consider the linear data below. Write an equation for H in terms of x and then graph the line.

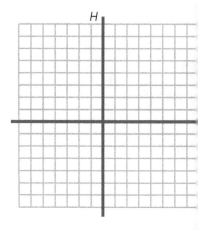

x	4	8	14	...
H	−1	−7	−16	...

133. How can you confirm that $(-20, 30)$ is located on the line in the previous scenario? Describe your method in words, and then verify your method by showing whether or not $(-20, 30)$ is on the line.

134. A candle is lit in the morning and slowly burns throughout the day. Its weight decreases at a constant rate as it burns. After the candle has been burning for 4 hours, its weight is 11 ounces. After it has been burning for 12 hours, its weight is 9 ounces.

    a. Write an equation that represents the weight of the candle, W, after it has been burning for h hours.

    b. How much will the candle weigh after it has been burning for 24 hours?

135. You have worked with linear equations in two different forms: Standard Form [Ax + By = C] and Slope-Intercept Form [y = mx + b]. You will eventually learn more about a third structure, which is known as Point-Slope Form [y – y_1 = m(x – x_1)]. Use what you have learned so far to convert the following equation to Slope-Intercept Form.

$$y-3=-\frac{1}{3}(x+6)$$

136. Rewrite each equation to convert it to Slope-Intercept Form.

   a. $y+5=-\frac{3}{4}(x-4)$

   b. $y-1=-(x-8)$

137. Graph the equation $y+2=\frac{3}{2}(x-2)$.

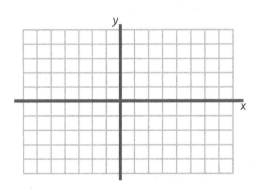

138. ★The graph to the right shows the linear relationship between temperatures measured in Celsius and Fahrenheit. Use the graph to answer the following questions.

   a. What is 0°C when measured in Fahrenheit?

   b. What is 0°F when measured in Celsius?

   c. What is 40°C when measured in Fahrenheit?

   d. What is –40°F when measured in Celsius?

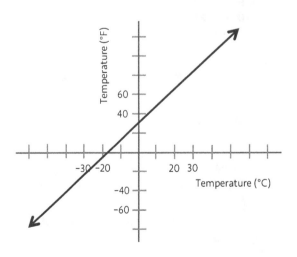

139. Consider the ordered pairs $(-5, -2)$ and $(4, -5)$.

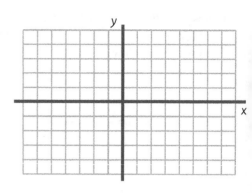

    a. Write the equation of the line that passes through the given ordered pairs, using Slope–Intercept Form.

    b. Graph the line.

140. Plot a point that has a *y*-value of 4. Now plot another point that has a *y*-value of 4. Pick three more points such that each has a *y*-value of 4. Draw a line that passes through all of your points. What do you notice about your line? Try to figure out the equation of this line.

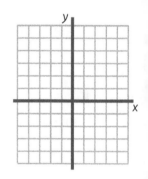

141. Plot a point that has an *x*-value of –5. Now plot another point that has an *x*-value of –5. Pick three more points such that each has an *x*-value of –5. Draw a line that passes through all of your points. What do you notice about your line? Try to figure out the equation of this line.

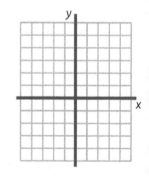

142. Two lines are shown on the Cartesian Plane to the right.

    a. What is the equation of the solid line?

    b. What is the equation of the dashed line?

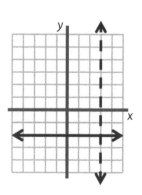

143. Graph each equation.

a. $x = -4$

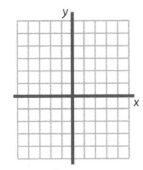

b. $y = -4$

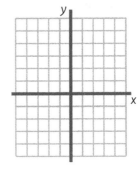

144. Graph each equation.

a. $y = 2$

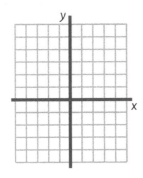

b. $x = 2$

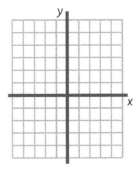

145. Consider the graph to the right. Rotate the image 90° in the clockwise direction. If the *x*-axis is always the axis that is horizontal (parallel to the ground), what is the slope of the line that you see after you rotate the image?

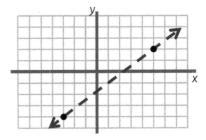

146. You and your friend are in a hotel and you want to go down to the first floor. There are stairs and an elevator, so you challenge each other to a race. Your friend takes the elevator and you take the stairs. You both start at the same time. The equations below model your two separate descents.

You: $H = -5t + 50$ Your friend: $H = -8t + 50$

In the two equations, *H* is the height above the first floor after you each have been descending for *t* seconds.

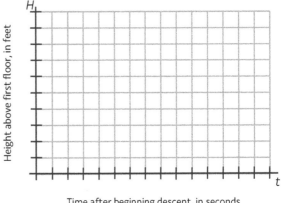

Time after beginning descent, in seconds

a. Graph both of the lines in the graph provided.

b. Who gets to the bottom first and by how many seconds does that person win the race?

147. Equations can be written in different forms. For example, start with an equation in Slope-Intercept Form, $y = -4x + 5$.

$$y = -4x + 5$$

    a. Move the term that contains x to the other side of the equation. Use the space to the right to show your work.

    b. Now arrange the terms to put the "x" term to the left of the "y" term.

After minimal effort, the equation has been converted to Standard Form: $Ax + By = C$.

148. Even if an equation in Slope-Intercept Form contains fractions, it can still be rearranged and written in Standard Form. Consider, for example, the equation $y = -\frac{3}{8}x + \frac{5}{8}$.

$$y = -\frac{3}{8}x + \frac{5}{8}$$

    a. Move the term that contains x to the other side. Use the space to the right to show your work.

    b. Write the "x" term to the left of the "y" term.

    c. Multiply both sides of the equation by ____. This will clear the fractions to make A, B and C integers.

Now the equation is in Standard Form: $Ax + By = C$. It is typical to eliminate the fractions, because the equation looks simpler when A, B, and C are integers.

149. Try to rewrite the equation $y = -\frac{3}{4}x - 7$ in Standard Form, where A, B, and C are integers.

150. Rewrite each equation to convert it to Standard Form.

    a. $y = -3x + 7$ b. $y = -\frac{1}{5}x + 6$ c. $y = \frac{3}{7}x - 2$

Use this page to record important ideas in the previous section or
for any other writing that helps you learn the topics in this book.

53

Section 9
LINEAR INEQUALITIES

The scenarios in this book so far have focused on linear equations such as $y = x + 4$. In this section, you will look at how the graph changes if you replace = with > or <.

151. In the Cartesian plane shown, plot all of the ordered pairs that have an x-value of 3. How many points can you find?

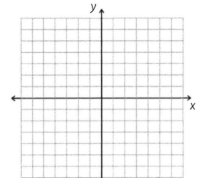

152. In the same plane to the right, graph all of the ordered pairs that have an x-value that is less than 3. How many points can you find?

When you combine the points that lie along the vertical line in your graph above with the points that occupy the space to the left of the vertical line, these points form the solution of the inequality $x \leq 3$. A common way to show this solution on a graph is to draw the vertical line for $x = 3$ and darken the region to the left of the line. The darkened region contains all of the ordered pairs with x-values less than or equal to 3.

153. In the Cartesian plane shown, plot all of the ordered pairs that have a y-value of –2.

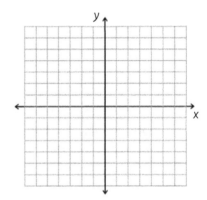

154. In the same plane shown to the right, graph all of the ordered pairs that have a y-value that is greater than –2.

155. When you combine the points on the horizontal line in the previous graph and the points that occupy the space above the horizontal line, these points form the solution of the inequality $y \geq -2$. How can you change the graph in the previous scenario to display the solution region for the inequality $y > -2$?

156. Graph each inequality.

a. $x > -4$

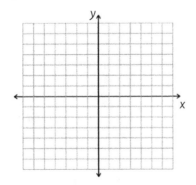

b. $y < 1$

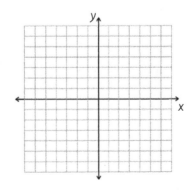

157. Graph each inequality.

a. $x \le 5$

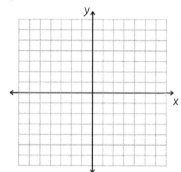

b. $-y \le -4$

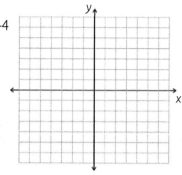

158. Circle the inequalities that would have a shaded region that includes the boundary line.

a. $y < x - 3$

b. $y \le 1 - 3x$

c. $y + x > 6$

d. $2x - 5y \le 10$

159. In the Cartesian plane shown, plot all of the ordered pairs that satisfy the relationship below.

The y-value is 1 more than the x-value.

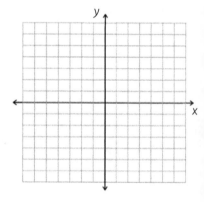

160. The points that you plotted to the right should form a line that can be represented by the equation $y = x + 1$. In the same plane shown to the right, plot all of the ordered pairs that satisfy the relationship $y > x + 1$.

In the previous scenario, when you combine the points <u>on</u> the slanted line in your graph and the points that are <u>above</u> that line, these points form the solution region for the inequality $y \ge x + 1$.

161. Match each inequality with its graph.

a. $y > -\dfrac{1}{4}x + 2$

b. $y \ge -2$

c. $y \le x - 3$

d. $x < -3$

i.

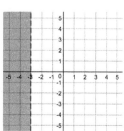

ii.

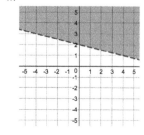

iii.

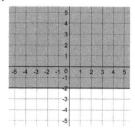

iv.

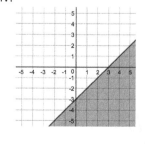

162. Graph the two equations shown below.

a. $y = -2x + 5$

b. $3x - 4y = 12$

163. Isolate the variable y in each inequality below.

a. $2x + y > 5$

b. $3x - 4y < 12$

164. Graph the two inequalities shown below.

a. $y < x - 3$

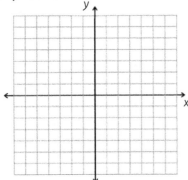

b. $2x - 5y \leq 10$

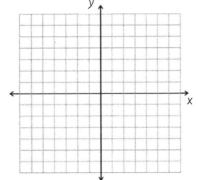

165. Is the ordered pair $(0, 0)$ part of the solution region in either graph in the previous scenario? Justify your answer using both the graph and the inequality.

57

166. Write the inequality that has the shaded solution region shown in each graph below.

a.
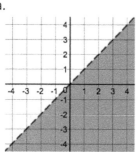

b.
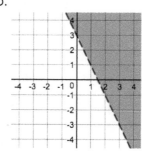

c.
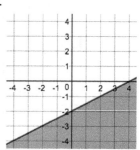

167. Graph the inequalities shown below.

a. $-y \le \dfrac{1}{4}x - 2$

b. $5x - 2y > 6$

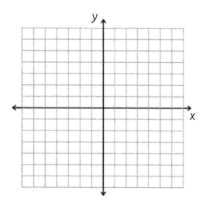

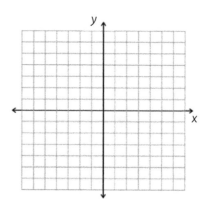

168. Which inequality has a solution region that covers a larger percent of the graph provided?

Inequality #1: $y \le -\dfrac{3}{7}x - 4$

Inequality #2: $y > -x + 5$

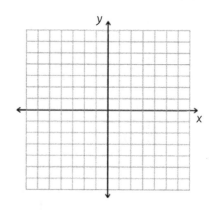

169. In the previous scenario, what percent of the graph is shaded by Inequality #2?

NOTES

Use this page to record important ideas in the previous section or
for any other writing that helps you learn the topics in this book.

Section 10
CUMULATIVE REVIEW

170. Audrey had a 12% discount off of her dinner bill of $57. What was the discount, in dollars?

171. If you drive on a toll road you must pay a flat fee of $0.25 plus an additional 9 cents per mile traveled. How far would you have to drive to make your toll at least $18?

172. The price of a pair of shoes is $63 after a 25% discount. What was the original price of the shoes?

173. If you have 4 equally weighted tests and earn a 65%, 78%, and 91% on the first three tests, respectively, what do you need to earn on the fourth test to achieve at least an 83% average?

174. In 2008, there were 122,295,345 people who voted during the U.S. Presidential election. In 2012, there were 131,313,820 voters. By what percent did the number of voters increase?

175. Sophia left a 20% tip for a meal. The total cost of the meal was $34.80. What was the cost of the meal before the tip was added?

176. ★At the end of the first semester, you would like to finish with an average of at least 92.5%. You have a 93% average after three assessments worth a total of 200 points. How many points do you need to earn on the final 80-point assessment in order to achieve your goal?

NOTES

Use this page to record important ideas in the previous section or
for any other writing that helps you learn the topics in this book.

Section 11
ANSWER KEY

1.	a. 15 seconds after 2:00pm, your heart rate is 75 beats per minute b. 105 seconds after 2:00pm, your heart rate is 150 beats per minute
2.	The temperature at 1pm is 5°F. At 3pm, it is 10°F. At 5pm, it is 15°F.
3.	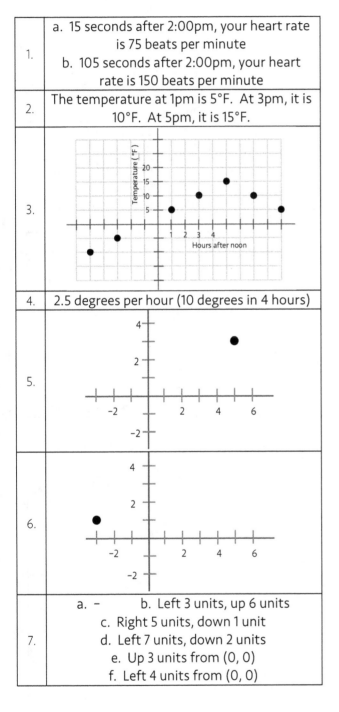
4.	2.5 degrees per hour (10 degrees in 4 hours)
5.	
6.	
7.	a. – b. Left 3 units, up 6 units c. Right 5 units, down 1 unit d. Left 7 units, down 2 units e. Up 3 units from (0, 0) f. Left 4 units from (0, 0)

	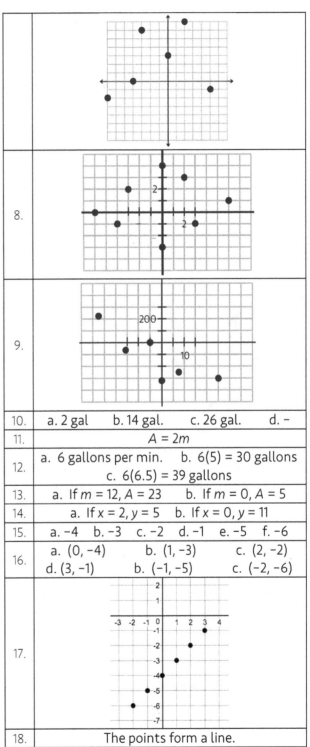
8.	
9.	
10.	a. 2 gal b. 14 gal. c. 26 gal. d. –
11.	$A = 2m$
12.	a. 6 gallons per min. b. 6(5) = 30 gallons c. 6(6.5) = 39 gallons
13.	a. If $m = 12$, $A = 23$ b. If $m = 0$, $A = 5$
14.	a. If $x = 2$, $y = 5$ b. If $x = 0$, $y = 11$
15.	a. –4 b. –3 c. –2 d. –1 e. –5 f. –6
16.	a. (0, –4) b. (1, –3) c. (2, –2) d. (3, –1) b. (–1, –5) c. (–2, –6)
17.	
18.	The points form a line.

63

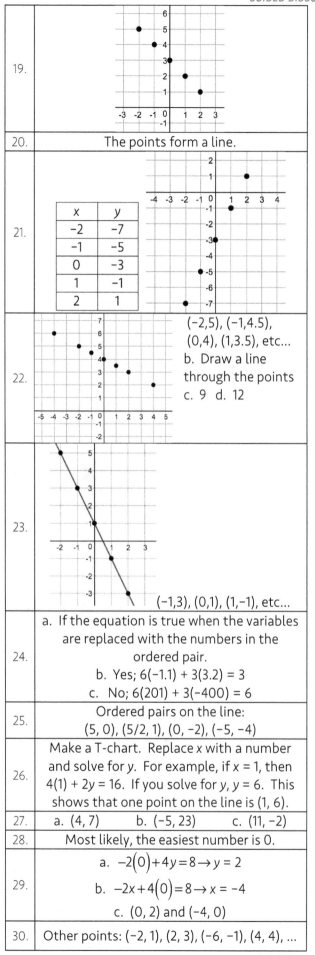

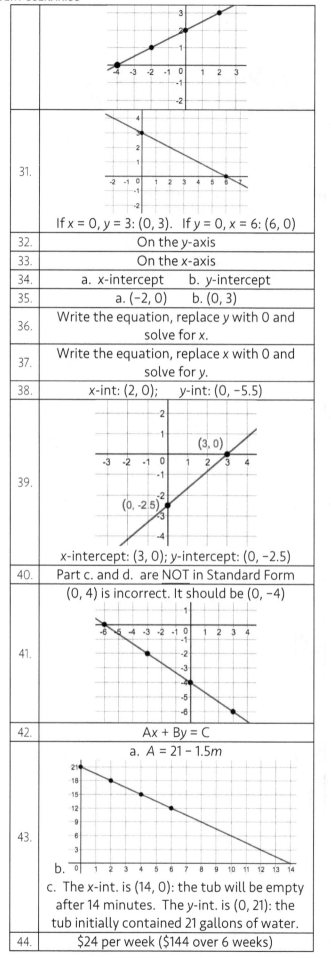

19.

20. The points form a line.

21.

x	y
−2	−7
−1	−5
0	−3
1	−1
2	1

22. (−2,5), (−1,4.5), (0,4), (1,3.5), etc…
b. Draw a line through the points
c. 9   d. 12

23. (−1,3), (0,1), (1,−1), etc…

24. a. If the equation is true when the variables are replaced with the numbers in the ordered pair.
b. Yes; 6(−1.1) + 3(3.2) = 3
c. No; 6(201) + 3(−400) = 6

25. Ordered pairs on the line:
(5, 0), (5/2, 1), (0, −2), (−5, −4)

26. Make a T-chart. Replace x with a number and solve for y. For example, if x = 1, then 4(1) + 2y = 16. If you solve for y, y = 6. This shows that one point on the line is (1, 6).

27. a. (4, 7)   b. (−5, 23)   c. (11, −2)

28. Most likely, the easiest number is 0.

29.
a. $-2(0) + 4y = 8 \rightarrow y = 2$
b. $-2x + 4(0) = 8 \rightarrow x = -4$
c. (0, 2) and (−4, 0)

30. Other points: (−2, 1), (2, 3), (−6, −1), (4, 4), …

31. If x = 0, y = 3: (0, 3).   If y = 0, x = 6: (6, 0)

32. On the y-axis

33. On the x-axis

34. a. x-intercept   b. y-intercept

35. a. (−2, 0)   b. (0, 3)

36. Write the equation, replace y with 0 and solve for x.

37. Write the equation, replace x with 0 and solve for y.

38. x-int: (2, 0);   y-int: (0, −5.5)

39. x-intercept: (3, 0); y-intercept: (0, −2.5)

40. Part c. and d. are NOT in Standard Form

41. (0, 4) is incorrect. It should be (0, −4)

42. Ax + By = C

43. a. A = 21 − 1.5m
b.
c. The x-int. is (14, 0): the tub will be empty after 14 minutes. The y-int. is (0, 21): the tub initially contained 21 gallons of water.

44. $24 per week ($144 over 6 weeks)

45.	a. The temp. is decreasing 2° per hour. b. $\frac{1}{2}$ hour per degree; it takes 30 minutes for the temp. to decrease by 1°
46.	a. \$4 per pound b. 0.25 pound
47.	a. The cost per liter b. (500, 40), (800, 100) c. Subtract the dollar amounts. Subtract the liters used. Divide to get \$ per liter. d. \$0.20 per liter
48.	a. $\frac{\$40}{400 \text{ texts}}$ or \$0.10 per text b. $\frac{400 \text{ texts}}{\$40}$ or 10 texts per dollar
49.	a. 2 ft b. $\frac{4.5}{54} = \frac{9}{108} = \frac{1}{12}$; Yes, it's legal.
50.	a. rises _6_ inches and runs _12_ inches b. $\frac{6}{12}$ c. $\frac{1}{2}$
51.	a. $\frac{8}{12} = \frac{2}{3}$ b. $\frac{10}{5} = \frac{2}{1}$ or 2 c. $\frac{50}{125} = \frac{2}{5}$
52.	a. The rise is 4. The run is 6. The slope is $\frac{2}{3}$. b. The rise is 3. The run is 5. The slope is $\frac{3}{5}$.
53.	a. rise: 8; run: 16; slope: $\frac{8}{16}$ or $\frac{1}{2}$ b. rise: 30; run: 100; slope: $\frac{30}{100}$ or $\frac{3}{10}$
54.	rise = 60; run = 300; $\frac{\text{rise}}{\text{run}} = \frac{60}{300} = \frac{1}{5}$; \$1 per 5 liters or \$0.20 per liter
55.	rise = 12; run = 6; $\frac{\text{rise}}{\text{run}} = \frac{12}{6} = \frac{2}{1}$; The temp. is decreasing by 2°F per hour.
56.	No. The temp. decreases each hr. in one graph, and increases per hr. in the other.
57.	a. down is negative b. left is negative
58.	a. up 3, right 5 or right 5, up 3 b. down 3, left 5 or left 5, down 3 c. $\frac{3}{5}$
59.	a. down 4, right 7 or right 7, down 4 b. up 4, left 7 or left 7, up 4 c. $-\frac{4}{7}$
60.	a. $\frac{\text{rise}}{\text{run}}$ b. $\frac{\text{rise}}{\text{run}}$
61.	Option 1: up 2, right 3 Option 2: down 2, left 3

62.	Option 1: down 2, right 3 Option 2: up 2, left 3
63.	 a. b.
64.	$\frac{2}{8}$ or $\frac{-2}{-8}$ or $\frac{1}{4}$ or $\frac{-1}{-4}$ Simplified, the slope is $\frac{1}{4}$.
65.	$\frac{-6}{9}$ or $\frac{6}{-9}$ or $\frac{10}{-15}$ or $\frac{-10}{15}$ Simplified, the slope is $-\frac{2}{3}$.
66.	a. solid: $-\frac{1}{3}$; dashed: 3 b. solid: $\frac{4}{3}$; dashed: $-\frac{3}{4}$
67.	The slope is $\frac{10-6}{8-2} \rightarrow \frac{4}{6} \rightarrow \frac{2}{3}$.
68.	The slope is $\frac{2-(-7)}{0-(-3)} \rightarrow \frac{9}{3} \rightarrow \frac{3}{1} \rightarrow 3$.
69.	a. $\frac{6-(-9)}{-3-3} \rightarrow \frac{15}{-6} \rightarrow \frac{5}{-2} \rightarrow -\frac{5}{2}$ b. $\frac{-8-2}{6-(-10)} \rightarrow \frac{-10}{16} \rightarrow -\frac{5}{8}$ c. $\frac{2-20}{\frac{1}{2}-\frac{7}{5}} \rightarrow \frac{-18}{-\frac{9}{10}} \rightarrow 20$
70.	rise = 0, run = nonzero number. The slope is $\frac{0}{\text{nonzero}}$ or 0.
71.	A horizontal line has a slope of 0, because it does not "rise." A fraction of $\frac{0}{\text{nonzero}}$ has a value of 0.

65

72.	rise = nonzero number, run = 0. The slope is $\frac{nonzero}{0}$ or undefined.
73.	A vertical line has an undefined slope, because it does not "run." A fraction of $\frac{nonzero}{0}$ is undefined.
74.	a. undefined b. 0
75.	a. 0 b. undefined
76.	Ax + By = C
77.	c, e. The line's slope is $\frac{3}{1}$ or **3**. d. The equation is $y = \underline{\mathbf{3}}x - 5$.
78.	c,e. The slope is $\frac{-2}{1} \rightarrow \underline{\mathbf{-2}}$. d. The equation is $y = \underline{\mathbf{-2}}x + 4$.
79.	The slope of the line is the same as the coefficient of "x" in the equation.
80.	The slope is 6.
81.	The slope is $\frac{3}{4}$.
82.	a. 4 b. $\frac{1}{2}$ c. 2 d. A
83.	$y = \frac{5}{11}x +$ anything
84.	$-5x + y = 10$ can be rewritten as $y = 5x + 10$, so the slope of the line is 5.
85.	$2x + 3y = 12$ can be rewritten as $y = -\frac{2}{3}x + 4$, so the slope of the line is $-\frac{2}{3}$.
86.	a. Replace x with 0 and solve for y. b. (0, –5) c. The y-intercept is (0, –5) and –5 is part of the equation.
87.	The y-intercept is (0, 4) and 4 is part of the equation.
88.	(0, –8)
89.	–31
90.	(0, 0)
91.	(0, 4)

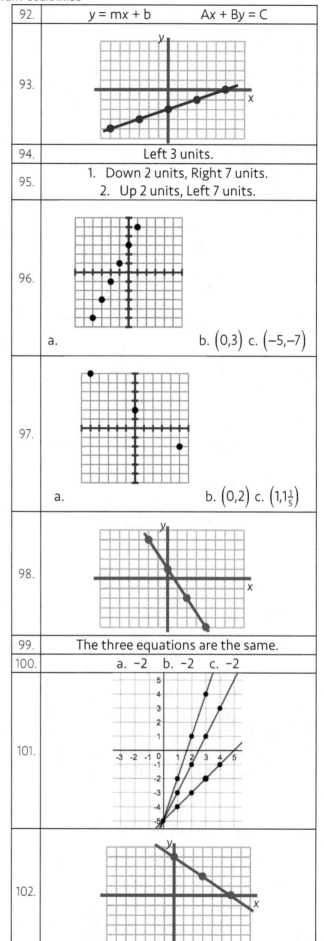

92.	$y = mx + b$ Ax + By = C
93.	
94.	Left 3 units.
95.	1. Down 2 units, Right 7 units. 2. Up 2 units, Left 7 units.
96.	a. b. $(0,3)$ c. $(-5,-7)$
97.	a. b. $(0,2)$ c. $\left(1,1\frac{1}{5}\right)$
98.	
99.	The three equations are the same.
100.	a. –2 b. –2 c. –2
101.	
102.	a.

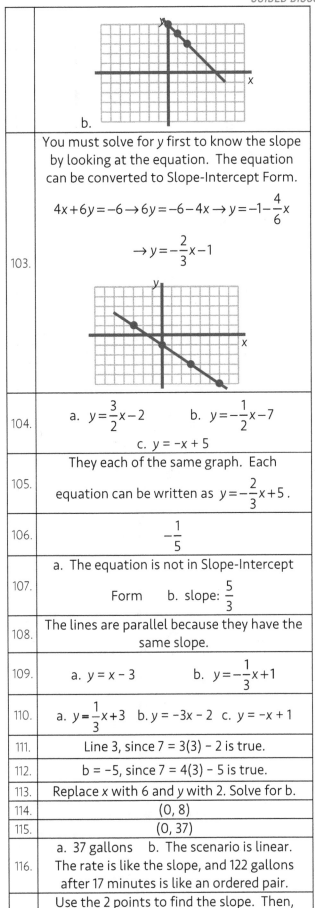

103. You must solve for y first to know the slope by looking at the equation. The equation can be converted to Slope-Intercept Form.

$$4x + 6y = -6 \rightarrow 6y = -6 - 4x \rightarrow y = -1 - \frac{4}{6}x$$

$$\rightarrow y = -\frac{2}{3}x - 1$$

104.
    a. $y = \frac{3}{2}x - 2$ b. $y = -\frac{1}{2}x - 7$

    c. $y = -x + 5$

105. They each of the same graph. Each

equation can be written as $y = -\frac{2}{3}x + 5$.

106. $-\frac{1}{5}$

107.
    a. The equation is not in Slope-Intercept

    Form   b. slope: $\frac{5}{3}$

108. The lines are parallel because they have the same slope.

109.
    a. $y = x - 3$ b. $y = -\frac{1}{3}x + 1$

110.
    a. $y = \frac{1}{3}x + 3$ b. $y = -3x - 2$ c. $y = -x + 1$

111. Line 3, since 7 = 3(3) – 2 is true.

112. b = –5, since 7 = 4(3) – 5 is true.

113. Replace x with 6 and y with 2. Solve for b.

114. (0, 8)

115. (0, 37)

116. a. 37 gallons   b. The scenario is linear. The rate is like the slope, and 122 gallons after 17 minutes is like an ordered pair.

117. Use the 2 points to find the slope. Then, insert one of the points into the equation and solve for b.

118.   a. $y = 4x + 3$ b. $y = -\frac{2}{3}x + 8$

119.   a. 7 gal./min.   b. 16 gallons

120. y = 7x + 16

121. $y = \frac{1}{4}x - \frac{1}{4}$

122.   a. $y = -\frac{4}{3}x + \frac{13}{3}$ b. $y = \frac{3}{4}x - 2$

123. perpendicular lines have slopes that are opposite reciprocals

124.
    a. dashed: $\frac{2}{3}$ solid: $-\frac{3}{2}$

    b. dashed: $-\frac{1}{3}$ solid: $\frac{3}{1}$ or 3

125. opposite reciprocals

126.   a. $-\frac{5}{2}$ b. $\frac{4}{7}$ c. –3   d. $\frac{1}{8}$

127.   a. $-\frac{5}{3}$ b. $\frac{1}{3}$ c. 7

128.   a. $\frac{3}{5}$ b. –3   c. $-\frac{1}{7}$

129. a. parallel   b. neither; the slopes are opposites but they are not reciprocals   c. perpendicular

130.
a. parallel lines

$$3x + 7y = 21 \rightarrow y = -\frac{3}{7}x + 3$$

$$-6x - 14y = 28 \rightarrow y = -\frac{3}{7}x - 2$$

b. neither; slopes are reciprocals but do not have opposite signs

$$30x + 40y = -400 \rightarrow y = -\frac{3}{4}x - 10$$

$$-4x - 3y = 12 \rightarrow y = -\frac{4}{3}x - 4$$

131.   a. A = 24w + 200   b. $680

132. $H = -\frac{3}{2}x + 5$

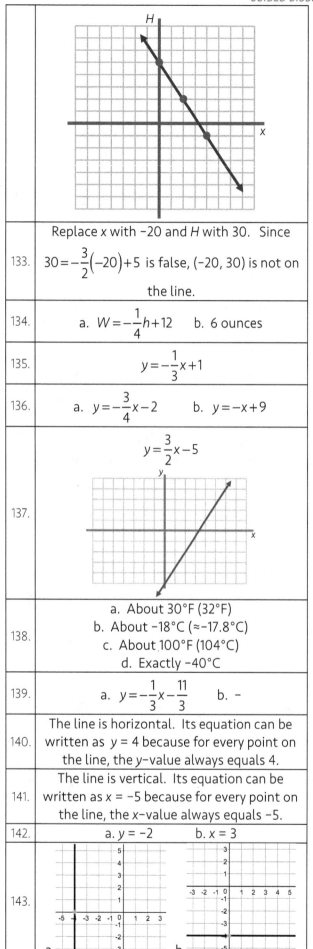

133.	Replace *x* with –20 and *H* with 30. Since $30 = -\frac{3}{2}(-20)+5$ is false, (–20, 30) is not on the line.
134.	a. $W = -\frac{1}{4}h+12$ b. 6 ounces
135.	$y = -\frac{1}{3}x+1$
136.	a. $y = -\frac{3}{4}x-2$ b. $y = -x+9$
137.	$y = \frac{3}{2}x-5$
138.	a. About 30°F (32°F) b. About –18°C (≈–17.8°C) c. About 100°F (104°C) d. Exactly –40°C
139.	a. $y = -\frac{1}{3}x - \frac{11}{3}$ b. –
140.	The line is horizontal. Its equation can be written as $y = 4$ because for every point on the line, the *y*-value always equals 4.
141.	The line is vertical. Its equation can be written as $x = -5$ because for every point on the line, the *x*-value always equals –5.
142.	a. $y = -2$ b. $x = 3$
143.	a. b.

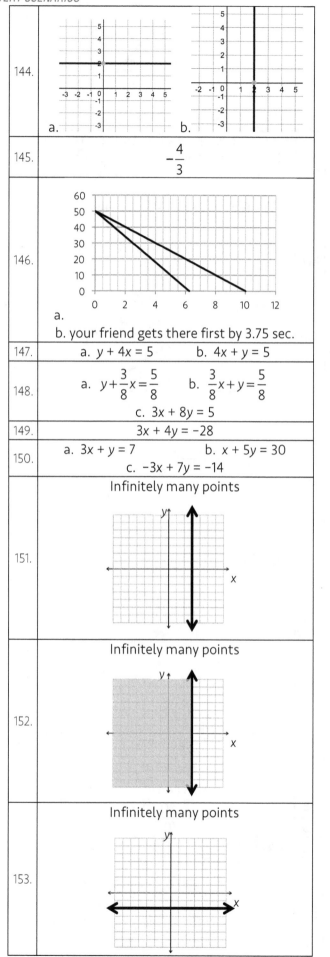

144.	a. b.
145.	$-\frac{4}{3}$
146.	a. b. your friend gets there first by 3.75 sec.
147.	a. $y + 4x = 5$ b. $4x + y = 5$
148.	a. $y + \frac{3}{8}x = \frac{5}{8}$ b. $\frac{3}{8}x + y = \frac{5}{8}$ c. $3x + 8y = 5$
149.	$3x + 4y = -28$
150.	a. $3x + y = 7$ b. $x + 5y = 30$ c. $-3x + 7y = -14$
151.	Infinitely many points
152.	Infinitely many points
153.	Infinitely many points

154. Infinitely many points

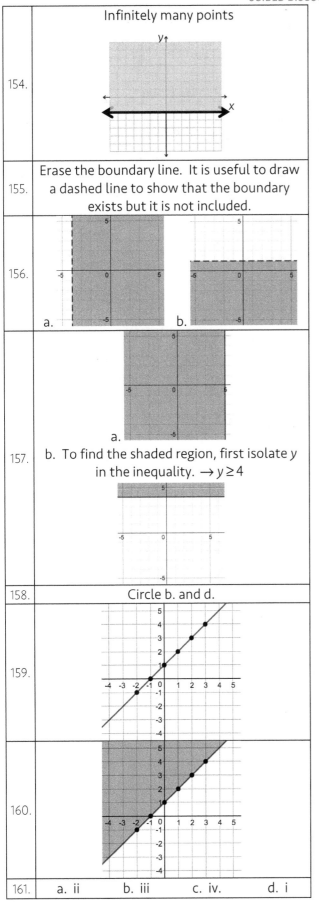

155. Erase the boundary line. It is useful to draw a dashed line to show that the boundary exists but it is not included.

156. a. b.

157. a.
b. To find the shaded region, first isolate y in the inequality. $\rightarrow y \ge 4$

158. Circle b. and d.

159.

160.

161. a. ii   b. iii   c. iv.   d. i

162.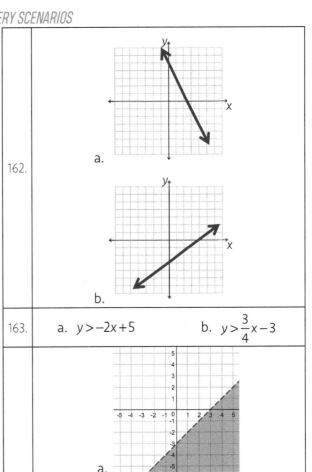
a.

b.

163. a. $y > -2x + 5$ b. $y > \dfrac{3}{4}x - 3$

164.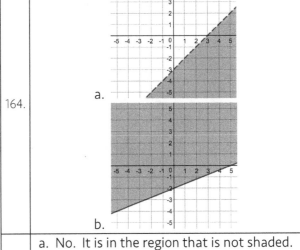
a.

b.

165. a. No. It is in the region that is not shaded. Using the inequality: $0 < 0 - 3$ is false.
b. Yes. It is in the shaded region. Using the inequality: $2(0) - 5(0) \le 10$ is true.

166. a. $y < x$ b. $y > -2x + 3$ c. $y \le \dfrac{1}{2}x - 2$

167. a. Isolate $y \rightarrow y \ge -\dfrac{1}{4}x + 2$. Reminder: dividing both sides by -1 changes the direction of the inequality.

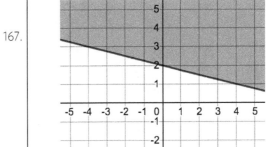

	b. Isolate $y \rightarrow y < \frac{5}{2}x - 3$ 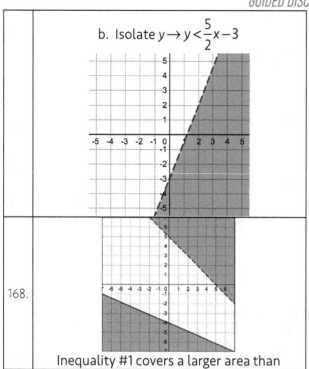
168.	Inequality #1 covers a larger area than

	Inequality #2 (42 units^2 versus 40.5 units^2)
169.	$\frac{40.5}{196} \rightarrow$ About 20.7%
170.	$6.84
171.	at least 198 miles (solve: $0.25 + 0.09m \geq 18$)
172.	$84
173.	at least a 98%
174.	About 7.37%
175.	$29.00
176.	73 out of 80 points

HOMEWORK & EXTRA PRACTICE SCENARIOS

As you complete scenarios in this part of the book, you will practice what you learned in the guided discovery sections. You will develop a greater proficiency with the vocabulary, symbols and concepts presented in this book. Practice will improve your ability to retain these ideas and skills over longer periods of time.

There is an Answer Key at the end of this part of the book. Check the Answer Key after every scenario to ensure that you are accurately practicing what you have learned. If you struggle to complete any scenarios, try to find someone who can guide you through them.

CONTENTS

Section 1
PLOTTING POINTS ON A GRAPH

1. The data points in the graph show how the temperature changed after school started one day.

   a. What information is contained in the point farthest to the left?

   b. What was the highest temperature observed, based on the data shown?

   c. Estimate the temperature at 1:00pm.

   d. What was the time when the temperature reached 68°F?

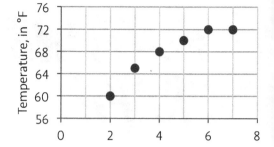

2. Describe the location of each ordered pair shown, and mark each ordered pair in the graph.

   DESCRIPTION: From the origin, move. . .

   a. $(-5, 2)$ left 5 units, up 2 units

   b. $(-1, -4)$

   c. $(3, 6)$

   d. $(4, -3)$

   e. $(-6, 0)$

   f. $(0, 5)$

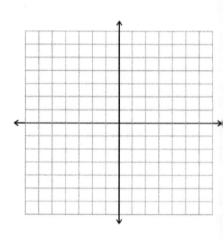

3. Estimate the location of each point below.

   a. $(-326, 49)$ b. $(174, -26)$

   c. $(0, -14)$ d. $(251, 35)$

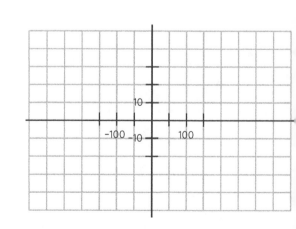

4. Which point is closer to the horizontal axis? Explain your choice.

Point #1: $(100, 4)$ Point #2: $(101, 3)$

5. Plot each point and use estimation to try to guess which point is closest to the location of $(0, 0)$.

a. $(2, 6)$ b. $(-3, 5)$ c. $(6, -2)$

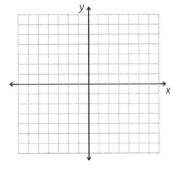

6. Which point is closer to the vertical axis? Explain your choice.

Point #1: $(-48, 27)$ Point #2: $(49, -26)$

7. In order to save enough money to buy a new jacket, you put $5 in a jar every day when you wake up. You start doing this on June 1st.

How much money have you saved at the end of each day listed below?

a. June 3rd

b. June 9th

c. June 15th

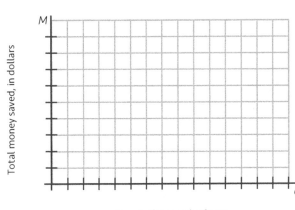

Days in the month of June

d. Draw points in the graph to show the total amount of money you have saved at the end of each day. Number the horizontal axis (*d*-axis) and vertical axis (*M*-axis) <u>before</u> you plot points.

8. For the previous scenario, write an equation that relates the amount of money you have saved, *M*, to the number of days, *d*, that you have been saving up your money.

9. Suppose you initially decided to set aside a different amount of money every day. If this change caused the equation in the previous scenario to become $M = 7.5d$, how much money would you have saved after 10 days?

10. Now change the equation to $M = 25 + \dfrac{5}{2}d$.

    a. What is the value of M when $d = 8$?

    b. What is the value of M when $d = 0$?

11. Consider the equation $y = 60 - 2x$.

    a. What is the value of y when $x = -2$?

    b. What is y when $x = 0$?

    c. For what value of x will $y = 100$?

Section 2

GRAPHING A LINE USING AN EQUATION & A T-CHART

12. Graph the equation $y = 3 - x$ by using a T-chart.

x	y
-2	$3-(-2) \rightarrow 5$
-1	
0	
1	
2	

ordered pair
$(-2, 5)$

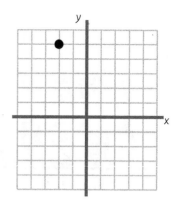

13. When you plot the points for the equation in the previous scenario, what do you notice about the way they appear in the graph?

14. Graph the equation $y = \frac{1}{2}x - 2$ using a T-chart. Find four ordered pairs using the chart. Then, graph one thousand more points in less than 5 seconds.

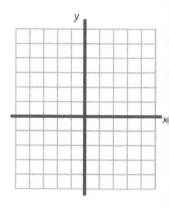

15. Graph $6x + 6y = 18$ using a T-chart. To save time, only graph 3 ordered pairs. Draw a line through your points to show that there are more points than the three you found.

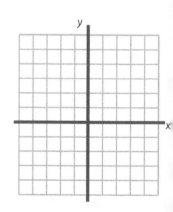

16. Consider the line in the previous scenario.

  a. Determine if $(2.5, 0.5)$ is on the line.

  b. Is $(-1.1, 4.2)$ on the line?

17. ★ Is $(1350, -1347)$ on the line formed by the equation $6x + 6y = 18$?

18. Graph all <u>six</u> of the points listed below.

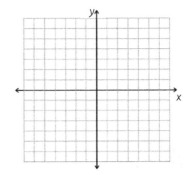

   a. $(6, 0)$ b. $(-1, 5)$ c. $\left(1, \dfrac{5}{3}\right)$

   d. $(-3, 3)$ e. $(0, -2)$ f. $(-6, 4)$

19. Some of the points shown in the previous graph are on the line formed by the equation $x + 3y = 6$. If you had to guess without doing any calculations, which points do you think are on this line?

20. The T-chart shown contains information that can be used to graph the equation $2x - 4y = 12$.

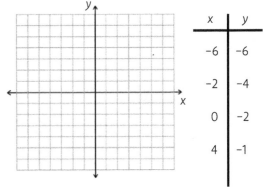

x	y
-6	-6
-2	-4
0	-2
4	-1

   a. How many points can be plotted using this T-chart?

   b. The T-chart has one error in it. Correct the error.

   c. Plot the points shown in the T-chart.

   d. Draw a line through the points to show that there are infinitely many points on the line formed by the equation $2x - 4y = 12$.

21. At what ordered pair does the line in the previous scenario cross the *y*-axis?

22. Where does the line in the previous scenario cross the *x*-axis?

23. It will sometimes be useful to write an equation in a different form. To practice this, isolate the variable y in the equation below.

$$2x + y = 5$$

24. Isolate the y in each equation shown.

    a. $6x + 2y = 8$ b. $12x - 8y = 40$

25. Graph the equation $y = -0.5x + 4$.

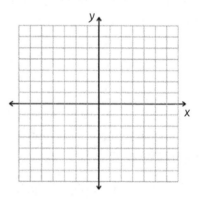

26. Graph the equation $30x - 10y = 50$.

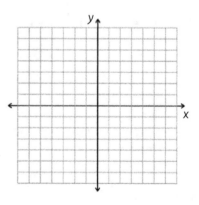

Section 3
GRAPHING A LINE USING ITS INTERCEPTS

83

27. How do you find the x- and y-intercepts of a line if you know the line's equation?

28. Consider the line represented by the equation 2x + 4y = 4.

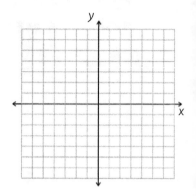

    a. Use the equation to find the y-intercept of the line.

    b. Find the x-intercept of the line.

    c. Graph these two points and draw a line through them.

    d. Find the coordinates of one more point on the line.

29. Graph 6x – 2y = 12 by following the strategy that you used in the previous scenario.

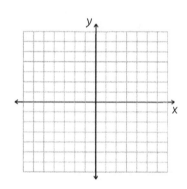

30. If a point has a y-value of 0, where will that point be located on the Cartesian plane?

31. If a point has an x-value of 0, where will it be located on the Cartesian plane?

32. Consider the graph shown to the right.

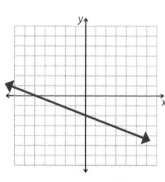

    a. Identify the x-intercept of the line.

    b. Identify the y-intercept of the line.

33. What is the x-intercept of the line shown?

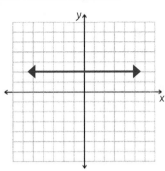

34. Explain why it is easy to get mixed up when you are trying to find x- and y-intercepts.

35. Identify the x- and y-intercepts of each equation shown.

   a. $4x - y = 8$

   b. $y = \dfrac{2}{5}x - 4$

   c. $-4x + 6y = 18$

36. Find the coordinates of the x- and y-intercepts of the equation $2x + 5y = -10$.
    Find the coordinates of one more point on the line and then graph the line.

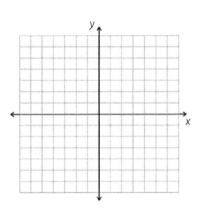

37. Write the Standard Form for a linear equation.

38. Circle the equations that are in Standard Form.

   a. $11x - 2y = 20$

   b. $y = -\dfrac{1}{3}x + 12$

   c. $-20x + 60y = 30$

   d. $9x + 2y - 5 = 0$

   e. $2x - 3y = -14$

   f. $x = \dfrac{3}{2}y + 1$

39. In an earlier scenario, you saved your money to buy a jacket. Then you started setting aside money again because you realized you wanted some boots as well. At the end of the day on November 2, you have $54. And at the end of the day on November 3, you have $60. Assume that you set aside the same amount of money every morning when you wake up.

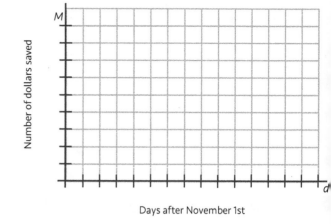

Days after November 1st

a. Write an equation that relates the amount of money you have saved, M, to the number of days, d, that have passed since November 1.

b. Graph the equation.

c. Identify the M-intercept of your graph and explain what it means in the context of this savings scenario.

40. Identify the x- and y-intercepts of the equation shown $-12x + 4y = 102$.

41. What is the y-intercept of the line shown?

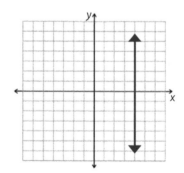

42. Consider the line shown. At what point will the line cross the x-axis?

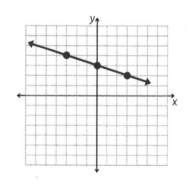

Section 4
CONSTANT RATES

43. If you are in the car on a long road trip and a GPS shows that you are 250 miles away from your destination at 5:00pm, you do not have enough information to figure out how fast the car has been moving. However, if you notice at 5:20pm that you are 228 miles from your destination, you can now estimate how fast the car has been moving since 5:00pm. How fast has the car been moving, in miles per hour?

44. Consider the graph to the right.

    a. What rate can be determined by analyzing the graph?

    b. What is the numerical value of the rate shown in the graph?

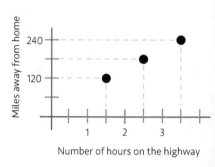

45. Let's take a look at another bathtub scenario. A portion of the graph is shown.

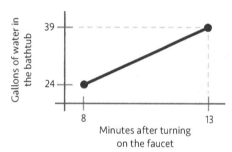

    a. Write the coordinates of the endpoints of the line segment.

    b. Explain how you can use the numbers in your ordered pairs to determine the flow rate of the bathtub faucet, in gal/min.

46. Two separate tanks are being emptied. The amount of water in tank #1 is shown by the dashed line. Tank #2 is the solid line.

    a. Which tank is emptying at a faster rate?

    b. Estimate the rate at which each tank is being emptied.

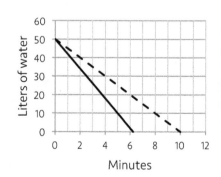

Section 5
THE SLOPE OF A LINE

47. Identify the slope of each ramp shown.

a.

0.5 in.

1 in.

b.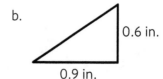

0.6 in.

0.9 in.

c.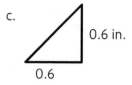

0.6 in.

0.6

48. Identify the slope of the ramp shown.

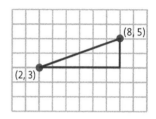

(8, 5)

(2, 3)

49. Identify the slope of the ramp shown.

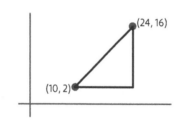

(24, 16)

(10, 2)

50. Identify the slope of the line shown and explain what the slope means using the labels for each axis.

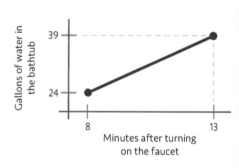

51. In the graphs below, pick a pair of points and determine the slope of the line. Then pick a different pair of points and find the slope of the line again. Compare your slopes.

a.

b.

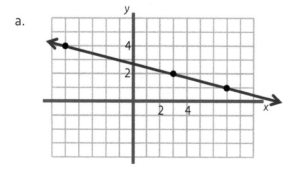

52. Identify the slope of the dashed line and the solid line in the graph below.

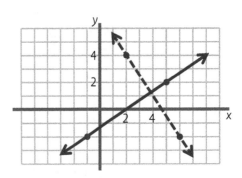

53. Draw the line that has the slope shown and passes through the point that is already plotted for you in the graph. Draw <u>three</u> more points before you draw the line.

a. Slope: 2

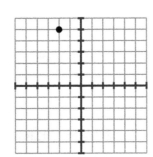

b. Slope: −3

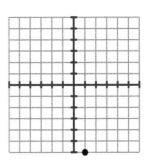

54. What is the slope of a line that passes through the ordered pairs $(-1,-10)$ and $(-6,5)$?

55. Determine the slope of the line passing through each set of ordered pairs.

a. $(-1, 10), (4, 5)$

b. $(-9, -3), (11, 7)$

56. Determine the slope of the line passing through each set of ordered pairs.

a. $(10, 132), (-50, 32)$

b. $(104, -12), (-6, -67)$

57. Mark two points on the dashed line to the right and use those points to find the slope of the line shown. Compare your slope with someone else. What do you notice?

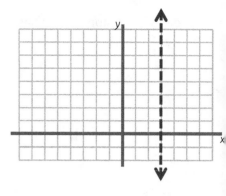

58. Mark two points on the dashed line to the right and use those points to find the slope of the line shown. Compare your slope with someone else. What do you notice?

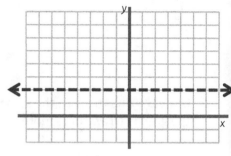

59. What is the slope of a vertical line? Why does a vertical line have this particular slope?

60. What is the slope of a horizontal line? Why does a horizontal line have this particular slope?

61. Two lines are shown on the Cartesian Plane to the right.

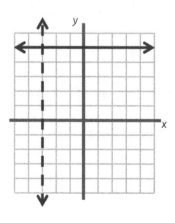

 a. What is the slope of the dashed line?

 b. What is the slope of the solid line?

62. Determine the slope of the line passing through each set of ordered pairs.

 a. $\left(7, -2\right), \left(-2, -2\right)$

 b. $\left(4, 9\right), \left(4, -5\right)$

63. ★Determine the slope of the line passing through each set of ordered pairs.

$$\left(3, \frac{2}{3}\right), (7, 6)$$

64. Identify the slope of the dashed line and the solid line in the graph below.

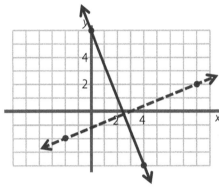

65. Determine the slope of the line passing through the ordered pairs.

$$(-7, 13) \text{ and } (14, 22)$$

66. A line with a slope of $-\dfrac{2}{3}$ passes through the ordered pairs below. What is the missing x-value in the second ordered pair?

$$(20, 11) \text{ and } (\underline{\hspace{0.5cm}}, 1)$$

67. Write the Standard Form of a linear equation.

Section 6

WRITING A LINE'S EQUATION IN SLOPE-INTERCEPT FORM

68. Consider the equation $y = \dfrac{1}{2}x - 3$.

   a. Fill in the T-chart and graph the ordered pairs.

   b. Plot four more points on the line.

   c. Use your graph to determine the slope of the line

   formed by the equation $y = \dfrac{1}{2}x - 3$.

x	y
-2	
0	
2	

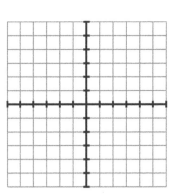

69. If the equation of a line is $y = 16x - 75$, what is the slope of this line?

70. If the equation of a line is $y = -\dfrac{11}{12}x + 100$, what is the slope of this line?

71. Fill in the blanks below.

   a. The line formed by the equation $y = -2x + 5$ has a slope of _____.

   b. The line formed by the equation $y = \dfrac{3}{7}x - 9$ has a slope of _____.

   c. The line formed by the equation $y = x - \dfrac{2}{9}$ has a slope of _____.

   d. The line formed by the equation $y = Mx + B$ has a slope of _____.

72. Write an equation for a line that has a slope of $-\dfrac{6}{7}$.

73. When a linear equation is arranged to look like $y = Ax + B$, the slope of the line will always be "A".
    What is the slope of the line formed by the equation $-7x + 2y = 14$?

74. What is the slope of the line that has an equation of 18x + 27y = 54?

75. What is the <u>y-intercept</u> of the line formed by the equation $y = -\dfrac{3}{4}x - 11$?

76. The line formed by the equation $y = \dfrac{1}{9}x + 57$ will cross the y-axis at (0, ____).

77. What are the coordinates of the y-intercept of the line formed by the equation $y = \dfrac{1}{9}x$?

78. When a linear equation is in the form y = Ax + B, the y-intercept will always be located at "B," or (0, B). What are the coordinates of the y-intercept of the line formed by the equation $-7x + 2y = 14$?

79. Write the Slope–Intercept Form of a linear equation.

80. Write the Standard Form of a linear equation.

81. Consider the equation $y = \dfrac{3}{4}x - 1$. Locate the y-intercept and draw a point to show where it is on the graph. Now identify the slope and use it to graph another ordered pair. Continue using what you know about the slope to graph other ordered pairs until you have 4 points on the line. Then quickly graph another seven thousand points.

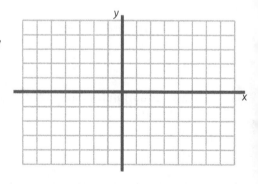

82. If the slope of a line is 1, you can start at one point on the line and move up 1 unit and right 1 unit to end up at another point on the line. If you start at a point on the line and move <u>down</u> 1 unit, how would you move horizontally to end up on the line again?

83. Suppose the slope of a line is −1. If you start at one point on the line, describe horizontal and vertical movements that would allow you to move to another point on the line.

84. Graph both equations on the same Cartesian Plane.

    a. $y = -x - 4$

    b. $y = x + 2$

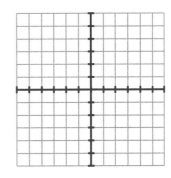

85. The slope of a line is −4. One of its points is shown in the graph.

    a. Draw three more points that are on this line.

    b. The y-intercept of this line is located at $\left(\underline{}, \underline{}\right)$.

    c. One of the points on this line is located at $\left(-5, \underline{}\right)$.

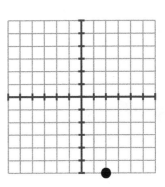

86. The slope of a line is $\dfrac{3}{4}$. One of its points is shown in the graph.

    a. Draw two more points that are on this line.

    b. The y-intercept of this line is located at $\left(\underline{}, \underline{}\right)$.

    ★c. One of the points on this line is located at $\left(3, \underline{}\right)$.

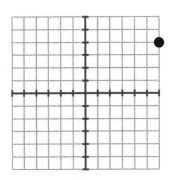

87. Graph the equation $y = -\dfrac{2}{5}x - 3$ by locating the y-intercept as your first point and then using the slope to find 2 more points.

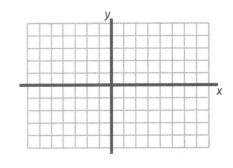

88. Graph the line given by the equation $y = -\dfrac{2}{3}x + 4$.

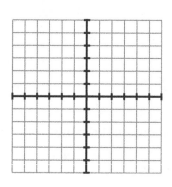

89. Circle the equation below that does <u>not</u> have the same graph as the other ones. Graph each equation on the same Cartesian Plane and compare the three graphs.

$$y = -\dfrac{3}{5}x + 1 \qquad y = \dfrac{-3}{5}x + 1 \qquad y = \dfrac{-3}{-5}x + 1$$

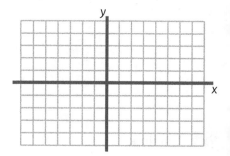

90. Try to quickly graph the equation $-x + 3y = 6$. What makes this more challenging than graphing the equations in the previous scenario?

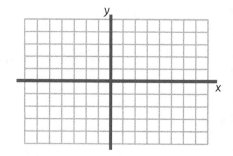

91. Graph the line given by each equation $-4x + 3y = 6$.

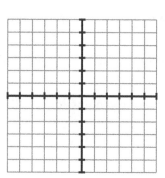

92. Each equation below is in Standard Form. Rewrite it to convert it to Slope-Intercept Form.

   a. $-7x - 3y = 9$ b. $9x + 4y = -12$ c. $5x - 5y = 30$

93. Start with an equation in Slope-Intercept Form. For example, start with $y = 3x + 7$.

    a. Move the term that contains x to the other side of the equation. Write your result.

    b. Now order the terms to put the "x" term to the left of the "y" term.

After minimal effort, the equation has been converted to Standard Form.

94. Even if an equation in Slope-Intercept Form contains fractions, it will still easily convert to Standard Form. Consider, for example, the equation $y = -\dfrac{1}{5}x + \dfrac{2}{3}$.

    a. Move the term that contains x to the other side. Write your result.

    b. Arrange the equation to place the "x" term to the left of the "y" term.

    c. Clear the fractions to make A, B and C integers. Write your result.

Now the equation is in Standard Form. It is <u>not actually necessary</u> to eliminate the fractions, but the equation looks less complex when A, B, and C are integers.

95. Rewrite each equation in Standard Form. Clear the fractions to make the coefficients integers.

    a. $y = -\dfrac{1}{2}x + 4$ b. $y = \dfrac{3}{5}x + 9$ c. $\dfrac{2}{9}x = y - 3$

96. Your math teacher writes the equation $2x - 7y = 14$ on the board and says the slope of that line is 2.

    a. Why is this the incorrect slope?

    b. Determine the slope of the line $2x - 7y = 14$.

97. Graph the equations $3x-4y=12$ and $8y-6x=-24$ on the same Cartesian plane. What do you notice?

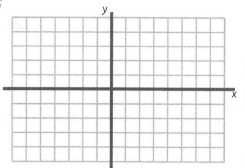

98. Graph the equations $4y+2x=8$ and $2x-y=-5$ on the same Cartesian plane. What do you notice?

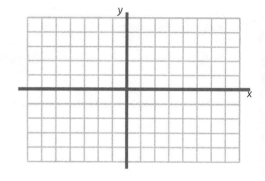

99. Determine the equation of each line shown below.

a.
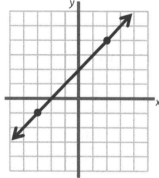

b.
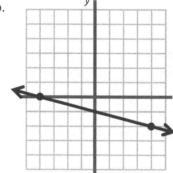

c.
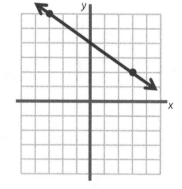

100. What is the y-intercept of a line that has a slope of $\frac{1}{3}$ and passes through the point $(6,\ 2)$?

101. What is the y-intercept of a line that has a slope of -3 and passes through the point $(90,\ 84)$?

102. While transferring a file to another device, you see that the file is being transferred at a rate of 3 megabytes per second. The size of the file is rather large so it is taking some time to send the file, but after 90 seconds, there are now only 84 MB left to transfer.

    a. How large was the original file, in megabytes?

    b. How is this scenario related to the previous scenario?

103. If you know 2 points that are on a line, how can you find the equation of the line in Slope-Intercept Form?

104. Find the equation of the line that passes through the given points.

    a. $(6, 8)$ and $(10, 10)$ b. $(3, -11)$ and $(-5, 5)$

105. After transferring the file in an earlier scenario, you started transferring another file. After 44 seconds, there are 801 MB left to transfer. After 80 seconds, there are still 720 MB left to transfer.

    a. How fast is the file being transferred, measured in megabytes per second?

    b. How large was the original file?

    ★c. How long does it take to transfer the entire file?

106. Find the equation of the line that passes through the points $(24, 846)$ and $(60, 765)$.

107. ★Determine the x-intercept of the line in the previous scenario. What does this value represent in the context of the previous file transfer scenario?

108. Find the equation of the line shown.

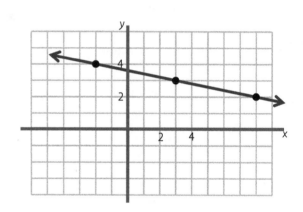

109. The following graph shows the amount of water in a pot as it sits over a fire and boils. Find the equation of the line in the graph, using the variables shown on each axis.

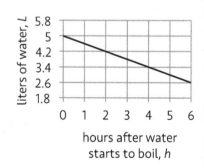

110. The equation of the line in the previous scenario contains information about that scenario. What details does the equation reveal about the water in the pot?

111. After school ends one Friday, you get in the car and travel to another state to spend the weekend away from home. The graph below displays information from a portion of your trip. Find the equation of the line in the graph, using the variables shown on each axis.

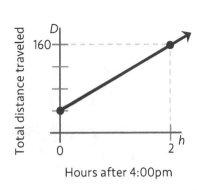

112. What details does the equation in the previous scenario reveal about the distance you have traveled?

113. The graph shows how the weight of a typical candle changes as it burns. Find the equation of the line in the graph, using the variables shown on each axis.

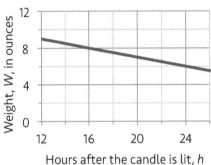

114. What details does the equation in the previous scenario reveal about the candle?

115. In the graph shown to the right, assume that ordered pairs that appear to be integers are integers.

a. Identify the equation of the solid line.

b. Identify the equation of the dashed line.

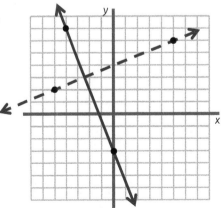

Section 7

PARALLEL & PERPENDICULAR LINES

116. The lines in the previous scenario are perpendicular. Notice their slopes. What must be true about the slopes of two lines to make those lines perpendicular?

117. Find the slope of each line shown.

a.

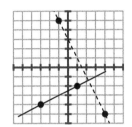

b.

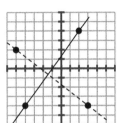

118. Both pairs of lines in the previous scenario are perpendicular. They intersect at right angles. Using what you have seen in the previous scenario, fill in the blanks below.

Perpendicular lines have slopes that are _____ _____ .

119. Write the opposite reciprocal of each number shown below.

a. $-\dfrac{5}{11}$

b. $\dfrac{9}{2}$

c. $-\dfrac{1}{7}$

d. 10

120. Fill in the box to make the two equations represent lines that are perpendicular.

a. $y=\dfrac{3}{2}x+1$
$y=\boxed{}x-4$

b. $y=\boxed{}x-8$
$y=0.5x+13$

c. $y=-\dfrac{1}{4}x+\dfrac{1}{3}$
$y=\boxed{}x+\dfrac{1}{2}$

121. Fill in the box to make the two equations represent lines that are parallel.

a.
$$y = \frac{3}{2}x + 1$$
$$y = \boxed{}x - 4$$

b.
$$y = \boxed{}x - 8$$
$$y = 5.28x + 13$$

c.
$$y = -\frac{1}{4}x + \frac{1}{3}$$
$$y = \boxed{}x + \frac{1}{2}$$

122. Do the equations represent parallel or perpendicular lines?

a.
$$y = -\frac{3}{5}x$$
$$y = \frac{5}{3}x + 2$$

b.
$$y = 6x - 5$$
$$y = -6x + 6$$

c.
$$y = x - 11$$
$$y = 11 - x$$

123. Do the equations represent parallel or perpendicular lines?

a.
$$8x - 2y = 4$$
$$x + 4y = 8$$

b.
$$5y + 55 = 4x$$
$$10y = 8x + 20$$

Section 8

SCENARIOS THAT INVOLVE LINEAR EQUATIONS

124. Consider the linear data below:

x	−9	−6	3	...
P	11	9	3	...

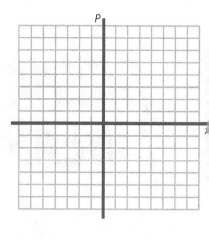

a. Write an equation for P in terms of x. Then graph the line.

b. Fill in the Blank (the options are below the sentence).

The point $(33, -16)$ is located _____ the line.

    i. On   ii. Close To   iii. Far Away From

125. As a plane approaches its destination, it begins descending at a constant rate. Four minutes after it begins its descent, its elevation is 28,000 feet. Nine minutes after it begins its descent, its elevation is 15,500 feet.

a. Write an equation that shows the relationship between the elevation, E, and the number of minutes, m, that the plane has been descending.

b. Is it reasonable to expect the plane to sustain its descent for 16 minutes?

126. Linear equations can be written in many different forms: Standard Form [Ax + By = C], Slope-Intercept Form [y = mx + b], and even Point–Slope Form [y − y₁ = m(x − x₁)]. Since you are not yet familiar with Point–Slope Form, rewrite the equations shown below to convert them to Slope-Intercept Form.

a. $y - 4 = \dfrac{3}{4}(x - 2)$

b. $y + 4 = -\dfrac{1}{2}(x - 10)$

127. ★Some movies are made using stop motion animation, which involves taking single photographs and then playing the sequence of photographs very quickly. By doing this, the objects captured in each photograph are made to look like they are moving. As you might imagine, it takes a very large number of photographs to make a stop motion film. Two film companies are shown below, with information about how many photographs they take to make films.

Film Company A	
Film time, in seconds	number of photographs
60	1,800
200	6,000

Film Company B
$T = 0.04p$

T is the length of the film, in seconds
p is the number of photographs in the film

a. How long is a film that is made by Company A if it contains 12,000 photographs? Write your result in minutes.

b. How long is a film that is made by Company B if it contains 15,000 photographs? Write your result in minutes.

c. Which company uses more photographs per second?

d. How many photographs per second does each company use?

128. Consider the ordered pairs $(5, 1)$ and $(-4, -5)$.

a. Determine the equation of the line that passes through the given ordered pairs. Write the equation in Slope–Intercept Form.

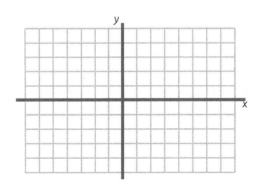

b. Graph the line.

129. Plot a point that has an *x*-value of 2. Now plot another point that has an *x*-value of 2. Pick three more points such that each has an *x*-value of 2. Draw a line that passes through all of your points. What do you notice about your line? What is the equation of this line?

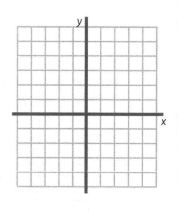

130. Plot a point that has a *y*-value of −3. Now plot another point that has a *y*-value of −3. Pick three more points such that each has a *y*-value of −3. Draw a line that passes through all of your points. What do you notice about your line? What is the equation of this line?

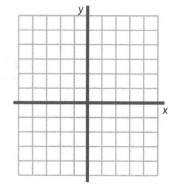

131. Two lines are shown on the Cartesian Plane to the right.

    a. Identify the equation of the solid line.

    b. Identify the equation of the dashed line.

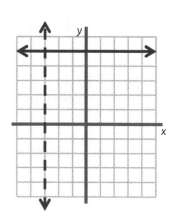

132. Graph the following three equations on the same plane.

    a. $x = 3$ b. $y = 4$

    c. $x = -4$ d. $y = -3$

    e. $y = -x$ f. $y = x + 1$

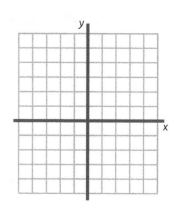

133. ★Consider the graph to the right.

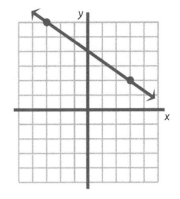

   a. If you rotate the graph 180°, what is the line's slope?

   b. If you rotate the original graph 90° in the clockwise direction, what is the line's equation?

   c. If you rotate the original graph 90° in the counterclockwise direction, what is the line's equation?

134. While riding in the car on the way to Philadelphia one day, you look out the window and notice that the car is passing mile marker 112. You look at the clock and notice that it is 10:15am. A little while later, after thinking about how slow the car seems to be moving, you look out the window again and notice mile marker 88. It's now 10:55am.

   a. How many miles have you driven?

   b. What was the average speed of the car over that 40-minute time period?

135. The points $(-2, -3)$, $(3, -3)$, and $(-2, 5)$ can serve as three vertices of a rectangle. Find the coordinates of the fourth vertex.

136. ★One afternoon in science class, your teacher allows you to shoot a rocket into the air. The rocket is launched from a platform on the ground. The graph shows how the rocket's height changes after it is launched.

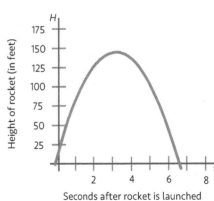

   a. Identify the x-intercepts of the graph and explain what they mean in the context of this scenario.

   b. Identify the y-intercept of the graph and explain what it means in the context of this scenario.

111

© Alex Joujan, 2020

137. While driving away from a gas station with your uncle one day, you feel too lazy to turn around and look at the latest gas prices. "Hey Uncle Sam, how much does gas cost lately?" Well, he's not one to hand out information easily so he says, "I just bought 4.5 more gallons of gas than the car beside me and my bill was $14.40 more than theirs."

    a. Determine the cost of gas and express your answer in dollars per gallon.

    b. Find the reciprocal of your rate in part a. and explain what it means.

138. Two middle school students make it to the finals of a math contest. The contest measures quickness in performing simple calculations. The final test contains 100 questions. If we assume the students worked at relatively constant rates, which student will probably win the contest? How do you know?

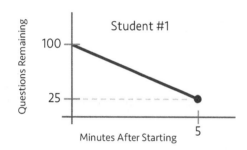 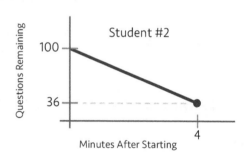

139. The Nathan's Hot Dog Eating Contest is held every year on July 4. Joey Chestnut won the contest every year from 2007 to 2013. The 3 graphs below represent Joey's results in 2007, 2010, and 2012, but they are out of order. If he got faster each year, match each graph with the year it represents.

a.

b.

c.

140. The graph of a circle is shown on the Cartesian Plane below.

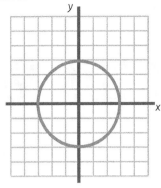

    a. Identify the *x*-intercept of the circle.

    b. Identify the *y*-intercept of the circle.

141. ★In the United States, temperature is usually measured in degrees Fahrenheit. In Canada, temperature is measured in degrees Celsius. This can get very confusing. Canadians believe that water boils at 100° while Americans believe that water boils at 212°. Canadians have found that water freezes at 0°, while Americans are quite certain that water freezes at 32°. The problem lies in the units of measurement.

    a. Use the given information to write an equation that relates the temperature in Fahrenheit, *F*, to the temperature in Celsius, *C*. Write your equation as $F = \ldots$

    b. If an American finds that Aluminum melts at 1220°F, at what temperature would a Canadian say that Aluminum melts?

    c. Rewrite your equation in part a. such that you isolate *C*.

142. Identify the rate shown in each graph and explain what it means in the context of the scenario.

a.

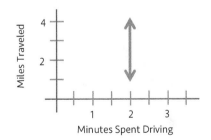

Minutes Spent Driving

b.

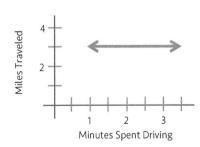

Minutes Spent Driving

Section 9
LINEAR INEQUALITIES

143. Graph each inequality.

a. $y > 3$

b. $x \le -4$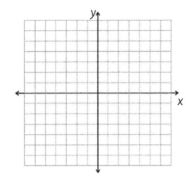

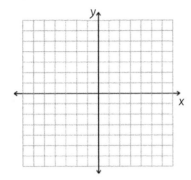

144. Graph each inequality.

a. $x > 2$

b. $y \le 0$

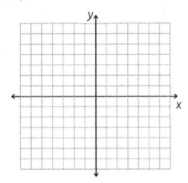

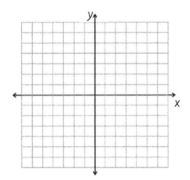

145. Circle each inequality that would have a shaded region that includes the boundary line.

a. $x + y \le 7$ b. $y > 5x + 2$ c. $-x + 4y \le 8$ d. $y > -2$

146. Match each inequality with its graph.

a. $y \le -\dfrac{2}{3}x + 2$ b. $y \ge -x + 3$ c. $y < -x - 3$ d. $y \ge -\dfrac{2}{3}x + 2$

i.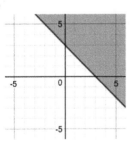

ii.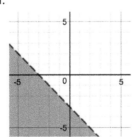

iii.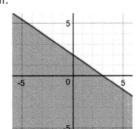

iv.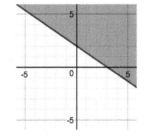

147. Isolate the variable y in the inequality below.

a. $3x + 5y < -30$

b. $-3y + 9x \geq 24$

148. Graph each inequality.

a. $y \geq -x + 2$

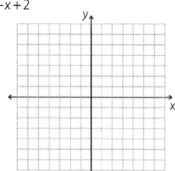

b. $x - 4y > 12$

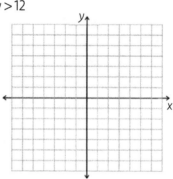

149. Write the inequality that has the solution set shown in each graph below.

a.

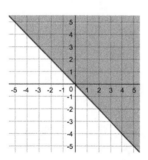

b.

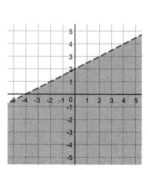

c.

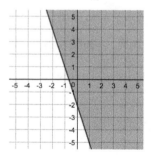

Section 10
CUMULATIVE REVIEW

150. Solve the equations shown below.

    a. $x - 8 = 3 - (x + 6)$ b. $\dfrac{2 + 3x - 5}{3} = 7$

151. A famous painting is sold in 2016 for $128,000. One year later, its value has increased by 12%. How much did the value of the painting rise, if you measure its rise in dollars?

152. Three games into the baseball season, the average attendance at the school's games was 3,528. After the fourth game, the total number of people who had attended the school's first 4 baseball games was 14,570. How many people were in attendance for the fourth game?

153. There were 122 million people who voted in the 2004 presidential election, which was 15.9% higher than the number of people who voted in 2000. How many people voted in the 2000 presidential election?

154. Identify the slope of each line segment shown. Write each slope as a fraction and then as a decimal. What do you notice?

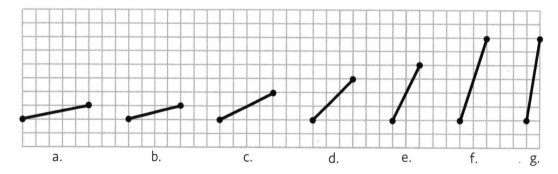

     a.   b.   c.   d.   e.   f.   . g.

Section 11
ANSWER KEY

1.	a. the temperature was 60°F at 10:00am b. 72°F c. 70°F d. 12:00noon
2.	b. left 1 unit, down 4 units c. right 3 units, up 6 units d. right 4 units, down 3 units e. left 6 units f. up 5 units
3.	
4.	Point #2 (the point with the smaller second number is closer to the horizontal axis)
5.	(−3, 5) is the closest to (0, 0)
6.	Point #1 (the point with the smaller first number is closer to the horizontal axis)
7.	a. $15 b. $45 c. $75

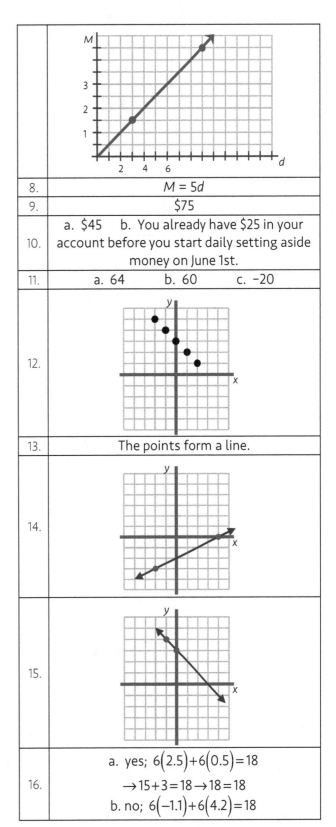

8.	$M = 5d$
9.	$75
10.	a. $45 b. You already have $25 in your account before you start daily setting aside money on June 1st.
11.	a. 64 b. 60 c. −20
12.	
13.	The points form a line.
14.	
15.	
16.	a. yes; $6(2.5)+6(0.5)=18$ $\rightarrow 15+3=18 \rightarrow 18=18$ b. no; $6(-1.1)+6(4.2)=18$

	$\rightarrow -6.6 + 25.2 = 18 \rightarrow 18.6 \neq 18$
17.	yes; $6(1350) + 6(-1347) = 18$ $\rightarrow 8100 - 8082 = 18 \rightarrow 18 = 18$
18.	
19.	$(6, 0)$, $\left(1, \dfrac{5}{3}\right)$, $(-3, 3)$, $(-6, 4)$
20.	a. 4 b. If $x = 0$, $y = -3$
21.	$(0, -3)$
22.	$(6, 0)$
23.	$y = 5 - 2x$ or $y = -2x + 5$
24.	a. $2y = 8 - 6x \rightarrow y = \dfrac{8 - 6x}{2} \rightarrow y = 4 - 3x$ b. $-8y = 40 - 12x \rightarrow y = \dfrac{40 - 12x}{-8}$ $\rightarrow y = -5 + \dfrac{3}{2}x$ or $y = \dfrac{3}{2}x - 5$
25.	
26.	$30x - 10y = 50 \rightarrow y = 3x - 5$

27.	To find the x-intercept, write the equation, replace the y-value with 0 and solve for x. To find the y-intercept, write the equation, replace the x-value with 0 and solve for y.
28.	a. When you plug in 0 for x, the y-value is 1, so the y-intercept is (0, 1). b. When you plug in 0 for y, the x-value is 2, so the x-intercept is (2, 0). c. d. points may include $(-2, 2)$, $(4, -1)$, etc...
29.	
30.	x-axis
31.	y-axis
32.	a. $(-5, 0)$ b. $(0, -2)$
33.	The line does not have an x-intercept because it is parallel to the x-axis so it will never cross the x-axis.
34.	It can be difficult to remember that replacing x with 0 allows you find the y-intercept and vice versa.
35.	a. x-int: $(2, 0)$, y-int: $(0, -8)$ b. x-int: $(10, 0)$, y-int: $(0, -4)$ c. x-int: $(-4.5, 0)$, y-int: $(0, 3)$
36.	
37.	$Ax + By = C$
38.	Circle a, c, and e.
39.	a. $M = 48 + 6d$ c. The M-intercept (0,48) is the amount of money in the account at the end of the day on November 1.

40.	x-int: (−8.5, 0) y-int: (0, 25.5)
41.	The line does not have a y-intercept because it is parallel to the y-axis so it will never cross the y-axis.
42.	(9,0); if you continue the pattern of the dots, you can put points at (6,1) and (9,0).
43.	The car traveled 22 miles in 20 minutes, which is 66 miles per hour.
44.	a. miles per hour b. 60 mph (120 miles over 2 hours)
45.	a. (8,24) and (13,39) b. Find how much the gallons change and how much the minutes change. Divide these values to find gal/min.
46.	a. Tank #2 b. Tank #1: 5 liters per minute Tank #2: 8 liters per minute
47.	a. $\dfrac{0.5}{1} \to \dfrac{1}{2}$ b. $\dfrac{0.6}{0.9} \to \dfrac{2}{3}$ c. $\dfrac{0.6}{0.6} \to 1$
48.	rise: 2; run: 6; slope: $\dfrac{2}{6} \to \dfrac{1}{3}$.
49.	rise: 14; run: 14; slope: $\dfrac{14}{14} \to \dfrac{1}{1} \to 1$
50.	Slope: 3; The flow rate of the bathtub faucet is 3 gal/min.
51.	a. $-\dfrac{1}{4}$ b. $\dfrac{3}{2}$
52.	dashed: $-\dfrac{3}{2}$; solid: $\dfrac{2}{3}$
53.	a. b.
54.	$\dfrac{5-(-10)}{-6-(-1)} \to \dfrac{15}{-5} \to -\dfrac{3}{1} \to -3$
55.	a. $\dfrac{5-10}{4-(-1)} \to \dfrac{-5}{5} \to -1$ b. $\dfrac{7-(-3)}{11-(-9)} \to \dfrac{10}{20} \to \dfrac{1}{2}$
56.	a. $\dfrac{32-132}{-50-10} \to \dfrac{-100}{-60} \to \dfrac{5}{3}$ b. $\dfrac{-67-(-12)}{-6-104} \to \dfrac{-55}{110} \to -\dfrac{1}{2}$
57.	rise = nonzero number, run = 0. The slope is $\dfrac{\text{nonzero}}{0}$ or undefined.
58.	rise = 0, run = nonzero number. The slope is $\dfrac{0}{\text{nonzero}}$ or 0.
59.	A vertical line has an undefined slope, because it does not "run." A fraction of $\dfrac{\text{nonzero}}{0}$ is undefined.
60.	A horizontal line has a slope of 0, because it does not "rise." A fraction of $\dfrac{0}{\text{nonzero}}$ has a value of 0.
61.	a. undefined b. 0
62.	a. $\dfrac{0}{-9} \to 0$ b. $\dfrac{-14}{0} \to$ undefined
63.	$\dfrac{5\frac{1}{3}}{4} \to \dfrac{\frac{16}{3}}{4} \to \dfrac{4}{3}$
64.	dashed: $\dfrac{2}{5}$; solid: $-\dfrac{5}{2}$
65.	$\dfrac{22-13}{14-(-7)} \to \dfrac{9}{21} \to \dfrac{3}{7}$
66.	The missing value is 35. $\dfrac{1-11}{x-20} = -\dfrac{2}{3} \to \dfrac{-10}{x-20} = \dfrac{-2}{3} \cdot \dfrac{5}{5} = \dfrac{-10}{15}$ $\dfrac{-10}{x-20} = \dfrac{-10}{15} \to x-20 = 15$ if $x = 35$
67.	Ax + By = C
68.	 a – b. c. slope: $\dfrac{1}{2}$
69.	slope: 16 or $\dfrac{16}{1}$
70.	slope: $-\dfrac{11}{12}$
71.	a. −2 b. $\dfrac{3}{7}$ c. 1 d. M
72.	$y = -\dfrac{6}{7}x +$ anything
73.	solve for $y \to y = \dfrac{7}{2}x + 7$; slope is $\dfrac{7}{2}$

74.	solve for $y \rightarrow y = -\frac{2}{3}x + 2$; slope is $-\frac{2}{3}$
75.	$(0, -11)$
76.	$(0, 57)$
77.	$(0, 0)$
78.	$(0, 7)$
79.	Slope-Intercept Form: $y = mx + b$
80.	Standard Form: $Ax + By = C$
81.	
82.	left 1 unit
83.	option 1: down 1 unit and right 1 unit option 2: up 1 unit and left 1 unit
84.	
85.	a. b. $(0,2)$ c. $(-5,22)$
86.	a. b. $(0,-0.5)$ c. $\left(3,1\frac{3}{4}\right)$

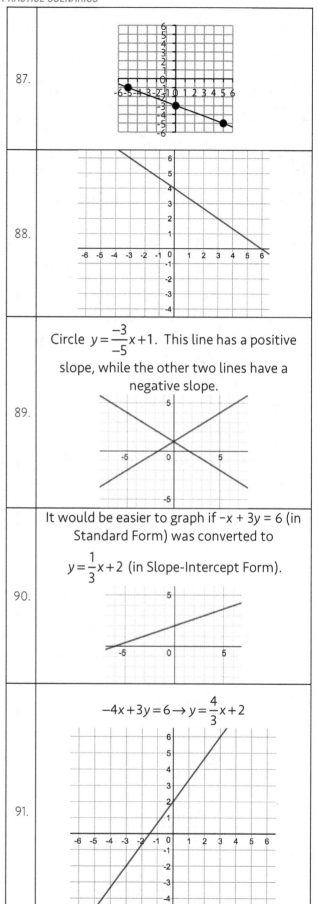

87.

88.

89. Circle $y = \frac{-3}{-5}x + 1$. This line has a positive slope, while the other two lines have a negative slope.

90. It would be easier to graph if $-x + 3y = 6$ (in Standard Form) was converted to $y = \frac{1}{3}x + 2$ (in Slope-Intercept Form).

91. $-4x + 3y = 6 \rightarrow y = \frac{4}{3}x + 2$

92.	a. $y=-\frac{7}{3}x-3$ b. $y=-\frac{9}{4}x-3$ c. $y = x - 6$
93.	a. $y - 3x = 7$ b. $-3x + y = 7$
94.	a. $y+\frac{1}{5}x=\frac{2}{3}$ b. $\frac{1}{5}x+y=\frac{2}{3}$ c. $3x + 15y = 10$
95.	a. $\frac{1}{2}x+y=4 \rightarrow x + 2y = 8$ b. $-\frac{3}{5}x+y=9 \rightarrow -3x + 5y = 45$ c. $\frac{2}{9}x-y=-3 \rightarrow 2x - 9y = -27$
96.	a. You must isolate y first to know the slope by looking at the equation. b. $y=\frac{2}{7}x-2 \rightarrow$ The slope is $\frac{2}{7}$.
97.	The equations represent the same line.
98.	The lines are perpendicular (they have opposite reciprocal slopes).
99.	a. $y = x + 2$ b. $y=-\frac{1}{4}x-1$ c. $y=-\frac{2}{3}x+4$
100.	In the equation, $y=\frac{1}{3}x+b$, replace x with 6 and y with 2. Solve for b. The y-intercept is $(0, 0)$.
101.	In the equation, $y = -3x + b$, replace x with 90 and y with 84. Solve for b. The y-intercept is $(0, 354)$.
102.	a. 354MB b. The slope in the prev. scenario is the transfer rate in this one, and the y-intercept in the prev. scenario is the original file size in this one.
103.	Use the 2 points to find the slope of the line. Then, insert one of the points into the equation $y = $ (slope)$x + b$ and solve

104.	for b. a. $y=\frac{1}{2}x+5$ b. $y = -2x - 5$
105.	a. 2.25 MB/sec b. 900 MB c. 400 seconds
106.	$y = -2.25x + 900$
107.	(400,0); this would be the time that it takes to transfer the entire file
108.	$y=-\frac{1}{5}x+3\frac{3}{5}$
109.	$L=-0.4h+5$
110.	The amount of water decreases by 0.4 liters every hour (the slope) and the pot was filled with 5 liters of water when it started to boil (the y-intercept).
111.	$D=60h+40$
112.	Your distance traveled increases by 60 miles every hour (the slope) and you have already traveled a distance of 40 miles at 4:00pm (the y-intercept).
113.	$W=-0.25h+12$
114.	The candle's weight decreases by one-fourth of an ounce every hour (the slope) and its original weight was 12 ounces when it was lit (the y-intercept).
115.	a. $y=-\frac{5}{2}x-3$ b. $y=\frac{2}{5}x+4$
116.	Two lines are perpendicular when their slopes are opposite reciprocals.
117.	a. solid: $\frac{1}{2}$ dashed: $\frac{-2}{1}$ or -2 b. solid: $\frac{4}{3}$ dashed: $-\frac{3}{4}$
118.	opposite reciprocals
119.	a. $\frac{11}{5}$ b. $-\frac{2}{9}$ c. 7 d. $-\frac{1}{10}$
120.	a. $-\frac{2}{3}$ b. $\frac{-2}{1}$ or -2 c. 4
121.	a. $\frac{3}{2}$ b. 5.28 c. $-\frac{1}{4}$
122.	a. perpendicular b. neither; the slopes are opposites but they are not reciprocals c. perpendicular; 1 and –1 are opposite reciprocals
123.	a. perpendicular lines $8x-2y=4 \rightarrow y=4x-2$ $x+4y=8 \rightarrow y=-\frac{1}{4}x+2$

	b. parallel lines $$5y+55=4x \rightarrow y=\frac{4}{5}x-11$$ $$10y=8x+20 \rightarrow y=\frac{4}{5}x+2$$
124.	a. $P=-\frac{2}{3}x+5$ b. Close To
125.	a. $E = -2{,}500m + 38{,}000$ b. No, the elevation would be –2,000ft at that point
126.	a. $y=\frac{3}{4}x+\frac{5}{2}$ b. $y=-\frac{1}{2}x+1$
127.	a. 6 minutes and 40 seconds b. 10 minutes c. Company A d. A: 30 per second; B: 25 per second
128.	$y=\frac{2}{3}x-\frac{7}{3}$
129.	Vertical line; equation is $x = 2$ because for every point on the line, the x–value always equals 2.
130.	Horizontal line; equation is $y = -3$ because for every point on the line, the y–value always equals –3.
131.	a. $y = 5$ b. $x = -3$
132.	
133.	a. $-\frac{2}{3}$ b. $y=\frac{3}{2}x-6$ c. $y=\frac{3}{2}x+6$
134.	a. $112 - 88 = 24$ miles b. $\dfrac{24\text{ miles}}{40\text{ minutes}}=\dfrac{24\text{ mi.}}{\frac{2}{3}\text{ hr}}=36$ mph
135.	(3, 5)
136.	a. x-intercepts are (0, 0) and (7, 0) since

	the rocket is launched off of the ground at 0 seconds and lands on the ground again after 7 seconds in the air. b. y-intercept is (0, 0). This shows that at the moment the rocket is launched, its height is 0 feet.
137.	a. $\dfrac{\$14.40}{4.5\text{ gal.}}=\$3.20$ per gallon b. $\dfrac{4.5\text{ gal.}}{\$14.40}=\dfrac{5}{16}$ gallon per dollar
138.	Student #1: $\dfrac{75\text{ questions}}{5\text{ minutes}}=15$ questions/min Student #2: $\dfrac{64\text{ questions}}{4\text{ minutes}}=16$ questions/min Student #2 will probably win.
139.	a. 6.8 hot dogs/min. – 2012 b. 5.5 hot dogs/min. – 2010 c. 5.4 hot dogs/min. – 2007
140.	a. Two x-intercepts: (–3, 0), (3, 0) b. Two y-intercepts: (0, 3), (0, –3)
141.	a. $F=\frac{9}{5}C+32$ b. 660ºC c. $C=\frac{5}{9}(F-32)$ or $C=\frac{5}{9}F-17.\overline{7}$
142.	a. The rate is impossible to describe, because the graph shows a car moving without time changing. b. 0 miles per minute. The car is not moving.
143.	a. b.

144.	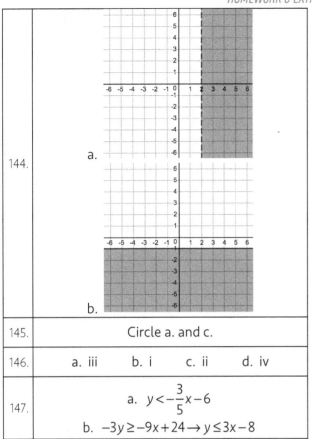 a. b.

145.	Circle a. and c.

146.	a. iii b. i c. ii d. iv

147.	a. $y < -\dfrac{3}{5}x - 6$ b. $-3y \geq -9x + 24 \rightarrow y \leq 3x - 8$

148.	a. 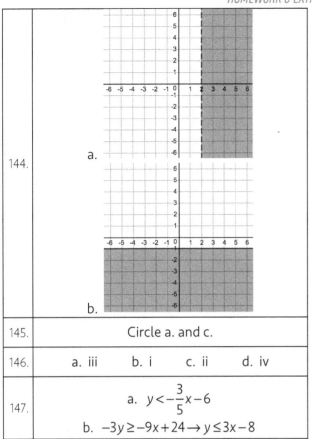 b.
149.	a. $y \geq -x$ b. $y < \dfrac{1}{2}x + 2$ c. $y \geq -3x - 2$
150.	a. $x = 2.5$ b. $x = 8$
151.	$15,360
152.	3,986
153.	Approx. 105 million
154.	a. 1/5 or 0.2 b. 1/4 or 0.25 c. 1/2 or 0.5 d. 1 e. 2 f. 3 g. 6 The slope gets larger as the line gets steeper.

CPSIA information can be obtained
at www.ICGtesting.com
Printed in the USA
LVHW051716140820
663222LV00008B/416

9 781713 283799